Marco Huber

Nonlinear Gaussian Filtering:
Theory, Algorithms, and Applications

Karlsruher Schriften zur Anthropomatik
Band 19
Herausgeber: Prof. Dr.-Ing. Jürgen Beyerer

Eine Übersicht aller bisher in dieser Schriftenreihe erschienenen Bände finden Sie am Ende des Buchs.

Nonlinear Gaussian Filtering: Theory, Algorithms, and Applications

by
Marco Huber

KIT Scientific Publishing

Habilitation, Karlsruher Institut für Technologie (KIT)
Fakultät für Informatik, 2015

Tag des Habilitationskolloquiums: 19. Januar 2015

Impressum

KIT Scientific Publishing

Karlsruher Institut für Technologie (KIT)
KIT Scientific Publishing
Straße am Forum 2
D-76131 Karlsruhe

KIT Scientific Publishing is a registered trademark of Karlsruhe Institute of Technology. Reprint using the book cover is not allowed.

www.ksp.kit.edu

Print on Demand 2015

ISSN 1863-6489
ISBN 978-3-7315-0338-5
DOI 10.5445/KSP/1000045491

Nonlinear Gaussian Filtering

— Theory, Algorithms, and Applications —

HABILITATIONSSCHRIFT

zur Erlangung der Venia Legendi
für das Fach Informatik

vorgelegt der Fakultät für Informatik
des Karlsruher Instituts für Technologie (KIT)

von

Dr.-Ing. Marco Huber

aus Kehl

Eingereicht am:	14.02.2014
Hauptgutachter:	Prof. Dr.-Ing. Jürgen Beyerer
Zweitgutachter:	Prof. Hugh Durrant-Whyte
	Prof. Peter K. Willett

Acknowledgement

While preparing this thesis, I have worked for three different organizations. Accordingly, I met many people that contributed to this work the one way or the other. I would like to spent the following lines to thank for the time, work, diversion, and inspiration invested by the contributors.

The first contributions to this thesis were made at the Intelligent Sensor-Actuator-Systems Laboratory (ISAS) of the Karlsruhe Institute of Technology (KIT). I would like to thank *Uwe D. Hanebeck*, director of ISAS, for supporting this thesis from the beginning. Also a big thank you to *Frederik Beutler* for his contributions to the semi-analytic filtering stuff and for running the experiments with the "Holodeck" as well as to *Peter Krauthausen* for his contributions to Gaussian mixture reduction. But most importantly: for their friendship.

When I moved on to the Fraunhofer Institute of Optronics, System Technologies, and Image Exploitation (IOSB), I met *Jürgen Beyerer*, who is the director of this institute. He actually motivated me for starting the whole habilitation endeavor and took the responsibility of the official habilitation process at the KIT. I can imaging that this work together with preparing a review of this thesis was consuming a significant share of time. So, many thanks for this.

At IOSB and the associated Vision and Fusion Laboratory (IES) I worked with many talented people. For the many fruitful discussions, barbecues, and for other thrilling activities I would like to thank *Christian Frese, Ioana Gheţa, Martin Grafmüller, Robin Gruna, Michael Heizmann, Sebastian Höfer, Christian Kühnert, Achim Kuwertz, Alexey Pak, Jennifer Sander, Michael Teutsch, Stefan Werling,* and last but not least *Philipp Woock.* Special thanks go to *Masoud Roschani.* Besides being a strong author of one of the papers used for this thesis, he acted and still acts as a very reliable contributor, organizer, and backup for my lecture "Probabilistische Planung". I would also like to thank *Gaby Gross,* who is by far more than a secretary for the IES. I enjoyed sharing work time and experiences about traveling and football with her.

The last three and a half years I worked for AGT International in Darmstadt, which was an interesting experience. I appreciate the time and space that was provided and blocked by both *Zachos Boufidis* and *Panayotis Kikiras.* They were great supporters of this thesis. I also have to thank *Florian Zeiger* for his intelligence, insights, and most importantly his friendship. Working at AGT would have been less fun without him. Finally, I would like to thank my students *Tobias Dencker* and *Jason Rambach* for their great work.

Outside these three organizations I would like to thank *Marc Deisenroth* for his ideas on Gaussian Processes and his bright mind in general. I'm also grateful for the review work done by *Hugh Durrant-Whyte* and *Peter Willett* and for the support that was given to this thesis by many computer science professors of the KIT. Here, I would like to mention *Rainer Stiefelhagen* and *Tamim Asfour* among many others.

Such a work not only needs to be backed from a professional side. Of at least the same importance is the support and understanding from beloved. I'm grateful for having *Karin* by my side. Life has become much more colorful with her and with our kids *Anna-Lena* and *Paul.*

Weinheim, January 2015 Marco Huber

To Anna-Lena and Paul

Contents

Notation

Conventions

x	Scalar
$\underline{x}$	Column vector
$\boldsymbol{x}$, $\underline{\boldsymbol{x}}$	Random variable/vector
$\hat{x}$, $\underline{\hat{x}}$	Realization of a random variable/vector
$\bar{x}$, $\underline{\bar{x}}$	Nominal point/vector
n_x	Dimension of vector $\underline{x}$
L_x, N_x	Number of elements of type x
$(\cdot)^p$	Predicted quantity (after prediction step)
$(\cdot)^e$	A posteriori quantity (after measurement update)
$(\cdot)^s$	Smoothed quantity (after smoothing)
$(\cdot)^x$, $(\cdot)_x$	Quantity related to variable/vector x
$(\cdot)_l$, $(\cdot)_n$	(Conditionally) linear and nonlinear substate
$(\cdot)_o$, $(\cdot)_u$	Observed and unobserved substate
$(\cdot)_k$	Quantity at discrete time step k
$(\cdot)_{0:k}$	Time sequence of quantities $\big((\cdot)_0, (\cdot)_1, \ldots, (\cdot)_k\big)$
$\tilde{(\cdot)}$	Approximate quantity
$(\cdot)^*$	Optimal solution of an optimization problem
$i : j$	Sequence of integers from i to j with $i < j$
$\cdot \mid \cdot$	Conditioning, i.e., the left-hand quantity is conditioned on the right-hand quantity

Conventions (Cont'd)

$\mathbf{A}$	Matrices are denoted by bold upper case letters
$\mathcal{A}$	Sets are denoted by calligraphic upper case letters
$g(\mathbf{X})$	Evaluation of function $g : \mathbb{R}^n \rightarrow \mathbb{R}$ for every column of matrix $\mathbf{X} = \left[\underline{x}_1\ \underline{x}_2\ \ldots \underline{x}_m\right]^{\mathrm{T}}$ with $\underline{x}_i \in \mathbb{R}^n$, i.e., the results corresponds to vector $\left[g(\underline{x}_1)\ g(\underline{x}_2)\ \ldots g(\underline{x}_m)\right]^{\mathrm{T}} \in \mathbb{R}^m$
$\mathbb{N}$	Natural numbers
$\mathbb{R}$	Real numbers
$\square$	End of theorem

Special Functions

$\mathcal{N}\left(\underline{x}; \underline{\mu}, \mathbf{C}\right)$	Multivariate Gaussian density function with mean vector $\underline{\mu}$ and covariance matrix $\mathbf{C}$
$\mathcal{N}\left(\underline{\mu}, \mathbf{C}\right)$	Multivariate Gaussian distribution
$\mathcal{GP}\left(\mu, \kappa\right)$	Gaussian process with mean function $\mu(\cdot)$ and covariance function $\kappa(\cdot,\cdot)$
$\delta\left(\underline{x}\right)$	Dirac delta distribution
$\delta_{i,j}$	Kronecker delta function
$D(.,.)$	Integrated squared difference (ISD)
$G(.\|.)$	Kullback-Leibler divergence (KLD)
$\mathrm{H}(.)$	Shannon/differential entropy
$\mathrm{I}(.,.)$	Mutual information

Operators

$\mathrm{diag}(\underline{x})$	Diagonal matrix with main diagonal elements according to $\underline{x}$
$\mathrm{diag}(\mathbf{A})$	Diagonal matrix with main diagonal elements taken from the main diagonal of $\mathbf{A}$
$\mathrm{vec}(\mathbf{A})$	Matrix vectorization, i.e., the columns of $\mathbf{A}$ are stacked to form a vector
$\mathbf{A}^{\mathrm{T}}$	Matrix transpose
$\mathbf{A}^{-1}$	Matrix inverse
$\mathrm{Tr}(\mathbf{A})$	Matrix trace
$\lvert\mathbf{A}\rvert$	Matrix determinant
$\lVert.\rVert_p$	p-Norm, where $p \in \mathbb{N}$. For $p = 2$, it corresponds to the L_2 or Euclidean norm
$\triangleq$	Definition
$\equiv$	Equivalent
$\preceq, \succeq$	Element-wise comparison of two vectors, i.e., $\underline{x} \preceq \underline{y}$ iff $x_i \leq y_i$ for all elements x_i, y_i of $\underline{x}$, $\underline{y}$
$*$	Convolution
$\dot{x}$	Time derivative of quantity x
$\otimes$	Kronecker product
$\odot$	Hadamard (element-wise) product
$\sim$	Distribution symbol
$\propto$	Proportional
$\mathrm{E}\{\cdot\}$	Expected value
$\mathrm{var}\{\cdot\}$	Variance
$\mathrm{Cov}\{\cdot\}$	Covariance
$\mathcal{O}(\cdot)$	Complexity class (Big-O according to Landau notation)

Reserved Symbols

$f(\cdot)$	General symbol for a probability density function
$\underline{g}(\cdot)$	General symbol for a nonlinear function
$\underline{a}_k(\cdot)$	System function
$\underline{h}_k(\cdot)$	Measurement function
$\underline{\boldsymbol{x}}$	System state
$\underline{\boldsymbol{z}}$	Measurement/observation
$\underline{\boldsymbol{w}}$	System noise
$\underline{\boldsymbol{v}}$	Measurement noise
$\underline{\boldsymbol{\theta}}$	(Hyper)parameter vector
$\underline{\mu}$	Mean vector
$\mathbf{C}$	Covariance matrix
$\mathbf{A}$	System matrix
$\mathbf{H}$	Measurement matrix
$\mathbf{K}$	Kalman gain matrix or kernel matrix
$\mathbf{J}$	Smoothing gain matrix
$\mathbf{I}_n$	Identity matrix of dimension $n \times n$
$\mathbf{0}_n$	Zero matrix of dimension $n \times n$
$\mathcal{X}_i$	Sample or sigma point
ω_i	Weighting coefficient
E_i	Non-central moment of order i
$\mathcal{D}$	Data set

Abbreviations

ADF	Assumed density filter
ADM	Atmospheric dispersion model
AGMF	Adaptive Gaussian mixture filter
CPKF	Chebyshev polynomial Kalman filter
CKF	Cubature Kalman filter
EKF	Extended Kalman filter
GPS	Global positioning system
GHKF	Gauss-Hermite Kalman filter
GP	Gaussian process
HPGF	Homotopic polynomial Gaussian filter
ISD	Integrated squared difference
KLD	Kullback-Leibler divergence
LRKF	Linear Regression Kalman filter
mae	Mean absolute error
MC	Monte Carlo
MCMC	Markov chain Monte Carlo
nees	Normalized estimation error square
nll	Negative log-likelihood
NN	Neural network
ODE	Ordinary differential equation
PF	Particle filter
PKF	Polynomial Kalman filter
QP	Quadratic program
RGP	Recursive Gaussian process regression
rmse	Root mean square error
RTSS	Rauch-Tung-Striebel smoother
SAGF	Semi-analytic Gaussian filter
SAGMF	Semi-analytic Gaussian mixture filter
SE	Squared exponential
SGMR	Superficial Gaussian mixture reduction
SIR	Sequential importance resampling

Abbreviations (Cont'd)

SOGP	Sparse on-line Gaussian process
SPGP	Sparse pseudo-input Gaussian process
SVM	Support vector machine
UKF	Unscented Kalman filter
UT	Unscented Transformation

Part I

Background & Summary

1

Introduction

In the early 1960s, the researchers of the NASA Apollo program faced severe problems with the navigation of their spacecraft. By observing external bodies like earth, moon, and stars, the pilot had to estimate the position, orientation, and velocity of the vehicle. These observations in combination with the high-dimensional models describing the dynamics of the spacecraft should allow correct guidance and trajectory following. At that point in time, the estimation toolbox merely contained algorithms like weighted least-squares and the Wiener filter. While the first one was computationally too complex for the on-board computer of the spacecraft, the latter was mathematically very involved and discrete-time measurements were not supported.

It was a lucky coincidence that at the same time Rudolf Kalman invented his famous recursive filtering approach, which is nowadays known as the Kalman filter. As Rudolf Kalman gave a presentation of his novel approach to NASA researchers, many of the open problems suddenly seemed to be solvable, even though a direct application of the Kalman filter was not possible—it was designed for linear problems, while spacecraft navigation is a nonlinear one. In addition, this novel filtering technique raised new questions like "How to deal with numerical instabilities or systematic measurement errors?", which are always present when it comes to a realization to practice. At the end the navigation problem was solved and the rest of the story is well-known history. For more details see [123].

The implementation of a modified version[1] of the Kalman filter for spacecraft navigation as part of the Apollo Guidance Computer was one of the first—if not the first—technical realization of *nonlinear Bayesian filtering* theory, which is the general topic of this thesis. Besides spacecraft navigation, nonlinear Bayesian filtering appears in many technical fields like robotics, control, telecommunications, signal processing, data fusion, or machine learning, where one wants to estimate the latent parameters or state[2] of a nonlinear dynamic system from noisy measurements and imperfect knowledge. It provides a general framework in which all appearing uncertainties are represented by means of probability distributions and the processing of these distributions occurs according to the calculus of probability theory[3].

1.1 Nonlinear Bayesian Filtering

The latent *system state* $\underline{x} \in \mathbb{R}^{n_x}$ comprises the smallest set of variables, which is necessary to completely describe the dynamic behavior of the considered system at any time instant. In order to estimate the latent state, three main components are required: First, a model of the system dynamics. Second, a model of the sensor. Both include statistical models of the noise processes affecting the system and the sensor. Third and finally,

1 Kalman-Schmidt filter, which is nowadays called the extended Kalman filter.

2 In this thesis, parameters and state are not distinguished if not stated explicitly. Thus, only the notion of a latent state is used from now on.

3 This is in contrast to *frequentist statistics*, where the probability of an event reflects the proportion of the event in an infinite number of trials. In the Bayesian viewpoint however, the probability describes the uncertainty of an event in a single trial [19, 160].

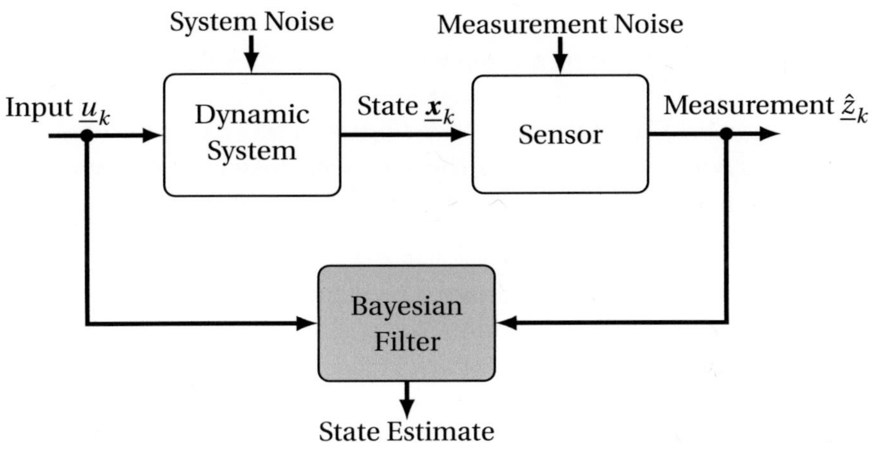

Figure 1.1: Block diagram of a dynamic system incorporating a Bayesian filter for estimating the latent state $\underline{\boldsymbol{x}}_k$.

the actual Bayesian filtering algorithm. The interactions of these three components are depicted in Figure 1.1 and mathematical formulations of the components are introduced in the following.

1.1.1 Dynamic Models and Measurement Models

The temporal evolution of the latent system state is typically described by means of differential equations in continuous time according to

$$\dot{\underline{\boldsymbol{x}}}(t) = \underline{\phi}\big(\underline{\boldsymbol{x}}(t), \underline{u}(t), \underline{\boldsymbol{w}}(t)\big) , \tag{1.1}$$

which models the system dynamics and thus, is denoted the *dynamic model*. Here, $\dot{\underline{\boldsymbol{x}}}(t)$ is the derivative of the system state $\underline{\boldsymbol{x}}(t)$ with respect to time. For realization on a computer, a discrete-time version of the dynamic model (1.1) is required, which is given by

$$\underline{\boldsymbol{x}}_{k+1} = \underline{a}_k\big(\underline{\boldsymbol{x}}_k, \underline{u}_k, \underline{\boldsymbol{w}}_k\big) , \tag{1.2}$$

where the random vector $\underline{\boldsymbol{x}}_k$ is the system state at discrete time step $k = 0,1,\ldots$ The time steps are related via $t_{k+1} = t_k + T$, where T is the sampling time. The state itself is represented by means of the probability density function $f_k^x(\underline{x}_k)$ at time step k. Furthermore, $\underline{a}_k(.)$ is the nonlinear system function, $\underline{u}_k$ is the vector of deterministic system inputs, and $\underline{\boldsymbol{w}}_k$ is the system noise. In the reminder of this thesis, only discrete-time models are considered.

As the state is assumed to be latent, i.e., it cannot be directly observed, measurements of a sensor are required to gather information about the system state. The *measurement model* in discrete-time is given by

$$\underline{\boldsymbol{z}}_k = \underline{h}_k\left(\underline{\boldsymbol{x}}_k, \underline{\boldsymbol{v}}_k\right) , \tag{1.3}$$

with nonlinear measurement function $\underline{h}_k(.)$, measurement noise $\underline{\boldsymbol{v}}_k$, and measurement $\underline{\boldsymbol{z}}_k$. An actual measurement $\hat{\underline{z}}_k$ of the sensor is a realization of $\underline{\boldsymbol{z}}_k$.

An important special case of the general models in (1.2) and (1.3) is the additive noise case. Here, the dynamic model and the measurement model are given by

$$\begin{aligned} \underline{\boldsymbol{x}}_{k+1} &= \underline{a}_k\left(\underline{\boldsymbol{x}}_k, \underline{u}_k\right) + \underline{\boldsymbol{w}}_k , \\ \underline{\boldsymbol{z}}_k &= \underline{h}_k\left(\underline{\boldsymbol{x}}_k\right) + \underline{\boldsymbol{v}}_k , \end{aligned} \tag{1.4}$$

respectively. This special case simplifies Bayesian filtering significantly, as the transition density and likelihood can be expressed analytically (see next section), but still it is sufficient for modeling a large class of important estimation problems [74].

The noise processes $\underline{\boldsymbol{w}}_k$ and $\underline{\boldsymbol{v}}_k$ in (1.2) and (1.3), respectively, account for typical uncertainties affecting the dynamic and measurement model. Examples are modeling uncertainties or external disturbances [74]. It is assumed that both noise processes are *white*. That is, both noise terms at time k are independent of the system state $\underline{\boldsymbol{x}}_k$ and independent of the noises at any other time step $n \neq k$. Both noise processes are presented by means of the probability density functions $f_k^w(\underline{w}_k)$ and $f_k^v(\underline{v}_k)$, respectively.

1.1.2 Recursive Filtering

Given a sequence of measurements $\hat{\underline{z}}_{0:n} = (\hat{\underline{z}}_0, \hat{\underline{z}}_1, \ldots, \hat{\underline{z}}_K)$, the estimation problem consists of determining an estimate of the system state $\underline{\boldsymbol{x}}_k$ based on $\hat{\underline{z}}_{0:K}$. Depending on the relation between the time steps k and K, three particular estimation tasks exist: if $k > K$, the estimation problem is called *prediction*, for $k = K$ it is called *filtering* or *measurement update*, and if $k < K$, it is called *smoothing*. Predictions and measurement updates are typically performed on-line, while smoothing is an off-line estimation task, as it aims for improving of past state estimates given additional information.

According to Bayesian filtering theory, the result of each of the three estimation tasks is given in form of a conditional probability density that represents the state estimate. Assuming that the system state $\underline{\boldsymbol{x}}_k$ is a Markov process[4], these densities can be calculated in a recursive fashion commencing from a initial state density $f_0^x(\underline{x}_0)$ at time step $k = 0$.

Prediction

In practical applications it is common that prediction and measurement update are performed alternatingly. Thus, without loss of generality, only the one-step prediction is considered here. Predictions over multiple time steps can be achieved by recursively performing one-step predictions.

Given the conditional density $f_k^e(\underline{x}_k) \triangleq f_k^x(\underline{x}_k | \hat{\underline{z}}_{0:k}, \underline{u}_{0:k-1})$ of the previous measurement update, the density of the predicted state for time step $k+1$

4 The state $\underline{\boldsymbol{x}}_k$ merely depends on the previous state $\underline{\boldsymbol{x}}_{k-1}$, but not on older states $\underline{\boldsymbol{x}}_l$ with $l < k-1$. This assumption automatically holds if the noise processes are white [93].

is calculated according to the so-called *Chapman-Kolmogorov equation* [93, 167]

$$f_{k+1}^p(\underline{x}_{k+1}) \triangleq f_{k+1}^x(\underline{x}_{k+1}|\underline{\hat{z}}_{0:k}, \underline{u}_{0:k}) = \int f(\underline{x}_{k+1}|\underline{x}_k, \underline{u}_k) \cdot f_k^e(\underline{x}_k)\,\mathrm{d}\underline{x}_k \,. \quad (1.5)$$

Here, the conditional density $f(\underline{x}_{k+1}|\underline{x}_k, \underline{u}_k)$ is the *transition density*, which depends on the dynamic model (1.2) and the system noise $\underline{\boldsymbol{w}}_k$. For the additive noise case (1.4), the transition density is given explicitly by

$$f(\underline{x}_{k+1}|\underline{x}_k, \underline{u}_k) = f_k^w(\underline{x}_{k+1} - \underline{a}_k(\underline{x}_k, \underline{u}_k)) \,, \quad (1.6)$$

where $f_k^w(\underline{w}_k)$ is the density of the system noise $\underline{\boldsymbol{w}}_k$.

Measurement Update

The measurement update incorporates the current measurement vector $\underline{\hat{z}}_k$ into the predicted density $f_k^p(\underline{x}_k)$. By employing Bayes' theorem, the measurement update results in the *posterior density* of the state given by

$$f_k^e(\underline{x}_k) = c_k \cdot f(\underline{\hat{z}}_k|\underline{x}_k) \cdot f_k^p(\underline{x}_k) \quad (1.7)$$

with normalization constant $c_k \triangleq 1/\int f(\underline{\hat{z}}_k|\underline{x}_k) \cdot f_k^p(\underline{x}_k)\,\mathrm{d}\underline{x}_k$. The term $f(\underline{\hat{z}}_k|\underline{x}_k)$ is known as the *likelihood* of the measurement $\underline{\hat{z}}_k$ and depends on the measurement model (1.3) and the measurement noise $\underline{\boldsymbol{v}}_k$. For the additive noise case (1.4), the likelihood can be expressed explicitly via

$$f(\underline{\hat{z}}_k|\underline{x}_k) = f_k^v(\underline{\hat{z}}_k - \underline{h}_k(\underline{x}_k)) \quad (1.8)$$

with $f_k^v(\underline{v}_k)$ being the density of the measurement noise $\underline{\boldsymbol{v}}_k$.

In [216] it is shown that the measurement update (1.7) is optimal from an information processing perspective, i.e., there is no loss or waste of information.

Smoothing

While the measurement update utilizes the current measurement for calculating the estimate, smoothing additionally incorporates future measurements. This restricts the application of smoothing for non-real time tasks, but the resulting estimate is more accurate as more information is used. The *smoothed* density $f_k^s(\underline{x}_k) \triangleq f_k^x(\underline{x}_k | \underline{\hat{z}}_{0:K}, \underline{u}_{0:K-1})$ of the state for any time step $k < K$ is calculated according the backward recursion[5]

$$f_k^s(\underline{x}_k) = f_k^e(\underline{x}_k) \cdot \int \frac{f(\underline{x}_{k+1} | \underline{x}_k, \underline{u}_k) \cdot f_{k+1}^s(\underline{x}_{k+1})}{f_{k+1}^p(\underline{x}_{k+1})} \, \mathrm{d}\underline{x}_{k+1} \tag{1.9}$$

commencing from the posterior density $f_K^s(\underline{x}_K) \equiv f_K^e(\underline{x}_K)$. All ingredients in (1.9) are already provided from the prediction and measurement update.

1.1.3 Closed-form Calculation

Estimating the latent state from a sequence of noisy measurements can be considered a statistical inverse problem (see for instance [42]), to which Bayesian filtering provides an optimal solution by means of (1.5), (1.7), and (1.9). The resulting conditional density functions of the system state form a basis for calculating further statistics like the mean vector or the covariance matrix, which are of practical use. The optimal solution, however, is of conceptional value only, mainly for two reasons. Although the recursive nature of the Bayesian filtering equations avoids calculating a joint distribution comprising the system states of all time steps, the resulting conditional densities cannot be represented by means of a finite number of parameters in general [26]. Furthermore, a closed-form solution of the filtering equations is not possible in general due to the involved integrals and multiplications of density functions.

5 This recursion actually holds for *fixed-interval* smoothing. Fixed-lag and fixed-point smoothing can be directly derived from it.

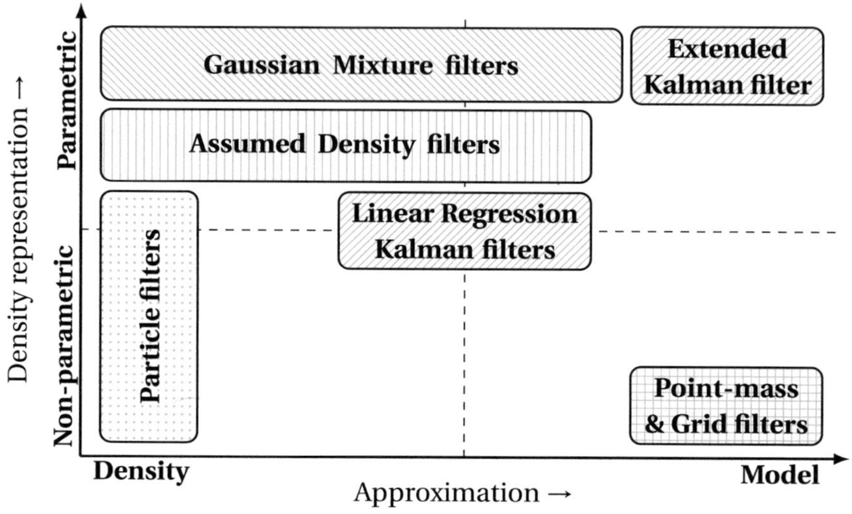

Figure 1.2: Popular approximate nonlinear filtering approaches.

Only for a few special cases, closed-form solutions can be found. An exception for instance exists for linear models affected by Gaussian noise, where the famous *Kalman filter* ([97] and Section 2.2.2) is optimal with closed-form expressions. In case of a finite state space, the *Wonham filter* [212] provides a closed-form solution. For general nonlinear models (1.2) and (1.3), however, an approximation of the optimal Bayesian solution is inevitable to obtain a feasible filter for practical applications. The following section gives a brief overview of major streams in approximate Bayesian filtering.

1.1.4 Approximate Filtering: State of the Art

In Figure 1.2, different groups of popular approximate filtering approaches are depicted. To show the differences between these groups, they are arranged regarding two approximation aspects:

1. The filtering approaches approximate the given nonlinear dynamic and measurement *models* or they approximate directly the *conditional densities* resulting from the optimal Bayesian filter.

2. For approximation purposes, a *parametric* (i.e., a fixed functional type) or *non-parametric* density representation is used.

Approximation of Models

One of mostly employed approximate nonlinear filtering approaches is the extended Kalman filter (EKF), which relies on a first-order Taylor-series expansion of the dynamic and measurement models [93, 174]. The Taylor-series expansion results in linear models, for which the Kalman filter equations can be employed. The major strengths of this approach are its simplicity and computational efficiency. Due to the linearization, the EKF is applicable only for mild nonlinearities. The EKF is described in more detail in Section 2.2.3.

Instead of a linearization, point-mass and grid filters [14, 33, 109, 173] utilize a discretization of the models, which leads to discrete version of Bayesian filtering problem. In doing so, all integrals in (1.5), (1.7), and (1.9) become summations, which are straightforward to evaluate. Point-mass and grid filters suffer from the so-called *curse of dimensionality* [16], as the number of discrete states increases exponentially with the dimension of the state space. Thus, these filtering methods are only feasible for low-dimensional problems.

Non-parametric Density Representation

Figure 1.2 indicates that the majority of the filters rely on an approximation of the density instead of the models. One explanation for this imbalance can be found in [96], where the authors state:

> *"It is easier to approximate a probability distribution than it is to approximate an arbitrary nonlinear function or transformation."*

Particle filters (PFs, see Appendix A) for instance utilize a weighted sample representation of the conditional densities—the so-called *particles*. In contrast to the point-mass and grid filters, where the discretization is performed deterministically, the particles are drawn randomly. This discrete density representation simplifies Bayesian filtering significantly, as prediction is simply performed by propagating the samples through the dynamic model (1.2), while measurement update and smoothing essentially boil down to adapting the particle weights. This simplicity is one of the key factors, why particle filters a very popular. Furthermore, PFs make no strong assumptions on the conditional densities. PFs form a Monte Carlo approximation, for which convergence towards the optimal Bayesian filtering solution with an increasing number of particles can be proven [59].

Although the convergence analysis is independent of the state dimension, practice shows that also PFs suffer from the curse of dimensionality [49]. This can be explained by the fact that with an growing state dimension also the volume to be filled with particles growth exponentially. An additional problem with PFs is sample depletion, i.e., over time most of the particles have zero-weight, which limits the representation of multimodel densities. As countermeasure resampling has to be employed. Also the choice of an appropriate proposal density from which the samples are drawn is critical.

In the recent years, modified PF algorithms have been proposed, which are not relying on a pure sample representation. To attenuate the aforementioned problems of PFs, these algorithms temporarily or even completely employ continuous densities like Gaussian densities [106], hybrid sample-Gaussian representations [201], or various mixture densities [1, 107, 129].

Parametric Density Representation

Thanks to their universal approximator property [122], *Gaussian mixture* densities are a welcome choice for approximating the conditional densities of the Bayesian filter. They provide an analytical and continuous representation that theoretically can approach the true density arbitrarily well, depending on the number of mixture components (see Section 3.1 for detailed introduction to Gaussian mixtures). Relevant statistics like mean or covariance can be derived in closed-form. In contrast to a single Gaussian density, determining the optimal parameters of a Gaussian mixture typically requires solving very demanding optimization problems. These optimization problems can be performed on-line in order to directly approximate the true conditional densities [76] or off-line, by replacing the transition density and likelihood by Gaussian mixtures [84, 87]. The latter case corresponds to the class of model approximating filters.

A more light-weight class of Gaussian mixture filters utilize multiple Gaussian filters like the EKF or linear regression Kalman filters (see next section) simultaneously, i.e., for each component of the mixture, a Gaussian filter is employed (see for example [7, 171] and Section 3). These filters combine the benefits of both worlds: the simplicity of Gaussian filters and the approximation power of Gaussian mixtures, but without sophisticated parameter optimization.

Alternatively to fitting the approximate density directly to the true density, *assumed density filters* (ADFs) calculate the approximate density in such a way that the moments of the true density are preserved—which is known as *moment matching*. This approach at least guarantees the correctness of some important statistics like mean and covariance, which might not be the case when directly approximating the density. Furthermore, for some nonlinear filtering problems the true conditional densities cannot be expressed analytically, but the moments are available in closed form (see for instance Section 2.5.2). For most filtering problems, however, calculating the desired moments requires numerical integration, which is computationally demanding especially for high-dimensional states.

Typical density representations employed in ADFs are Gaussians [67, 121], Edgeworth/Gram-Charlier series [39, 182, 183], or exponential densities [27, 29].

Linear Regression Kalman Filters

Basically, linear regression Kalman filters (LRKFs) like the famous unscented Kalman filter [96, 206] calculate a Gaussian approximation of the true conditional densities, whereas the parameters of the Gaussian density are obtained by propagating samples through the nonlinear models (1.2) and (1.3). In contrast to PFs, these samples are chosen in a deterministic fashion and capture the mean and covariance of the prior density exactly. Although this deterministic sampling clearly refers to a density approximation approach, there also exists an alternative interpretation [113, 114]: the same Gaussian density can be obtained by means of stochastic linearization of the models through the use of statistical linear regression (a theoretical treatment can be found in Section 2.2.5).

In contrast to the EKF, LRKFs provide more accurate estimates, while the computational complexity is almost the same. Furthermore, LRKFs are applicable to a larger class of filtering problems, as no differentiability of the system and measurement functions is required.

1.2 Research Topics

Especially with the advent of the LRKFs—besides the particle filters—significant improvements in approximating the optimal Bayesian filter have been achieved in the recent years. Motivated by the benefits of these Gaussian filters, in this thesis three major research topics grouped around *nonlinear Gaussian filtering* are covered. At first, further improvements of Gaussian filtering in general and LRKFs in particular are proposed. These improvements are then used for the two other research topics: filtering via Gaussian mixtures and Gaussian process models. In the following, for each of the three topics, particular reasons are identified why dealing with Gaussians, Gaussian mixtures, and Gaussian processes in the context of Bayesian filtering is reasonable and thus, worth for further investigation. Also limitations of state-of-the-art approaches are exposed, which are resolved in this thesis.

Why Gaussian Filtering?

Besides its optimality, a major reason of the wide application of the Kalman filter is its simplicity; all calculations are performed on the basis of matrix calculus (see Section 2.2.2). The EKF and LRKFs leverage the elegant Kalman filter equations for nonlinear filtering problems by assuming that the optimal conditional densities can be sufficiently well approximated by Gaussians. Additionally, these filters do not suffer from the curse of dimensionality as the number of parameters of a Gaussian density, i.e., the number elements of the mean vector and covariance matrix, merely grow quadratically with the state dimension. This is different for many other approaches like PFs. Gaussian filters are especially useful in applications where limited computational demand and memory usage is key, but an essential consideration of uncertainty is necessary.

Current Gaussian filters are typically applied in a black-box fashion, i.e., without a detailed analysis of the properties of the given dynamic and measurement models. Hence, approximations are applied even in cases, where at least some parts of the models do not require an approximate treatment. Furthermore, the Gaussian assumption is not only applied for the filtering results, it is also used for representing the joint distribution of state and measurement. This assumption is necessary for exploiting the Kalman filtering equations, but for many applications it is too strong and limits the quality of the approximation.

Why Gaussian Mixture Filtering?

Obviously, employing Gaussian densities to approximate the optimal Bayesian results may not be sufficient for every filtering problem[6]. Especially in cases, where the resulting densities are multimodal, heavily skewed, or have heavy tails, extending Gaussian filters towards using Gaussian mixture densities is desirable. Here, Gaussian filters benefit from the fact that this extension is straightforward to obtain as they can be applied on each mixture component independently. With an increasing number of components, it can be shown that Gaussian mixture filters utilizing EKF or LRKFs converge towards the optimal result [4].

One critical part of Gaussian mixture filters is the increase of the number of components. New components should only be introduced where the current approximation is not accurate enough. To limit the computation time and memory consumption, it is also necessary the reduce the number of components from time to time, especially when the number of components is disproportionate to the complexity of the density's shape.

Why Gaussian Process Filtering?

So far it was assumed that the dynamic and measurement models are known. This assumption does not hold in applications, where it is too complicated or even impossible to derive these models. These issues for instance may arise, when the underlying real system consists of many interacting elements like in robotics [131, 188], when the models cannot be calibrated sufficiently well like in WiFi-based localization [65] or in calibrating metal oxide sensors [125], or when the mapping between state and measurement is artificial like in classification problems [88]. Here, the mathematical models can be substituted by means of so-called *Gaussian process* (GP) models, which are non-parametric probabilistic models

6 In [191], it is illustrated that many natural and technical phenomena generate Gaussian distributions, but especially sociological systems—these also include financial systems—cannot be described with Gaussian statistics.

learned from data and which are a popular tool in machine learning (see Section 4.1 for a brief introduction to GPs).

Bayesian filtering with GP models so far has only been performed approximately. Furthermore, due to the non-parametric nature of GPs, the model accuracy but unfortunately also the model complexity increases with the size of data set used for learning. Performing Bayesian filtering can become infeasible if the data set grows over time.

1.3 Main Contributions

The research areas and questions posed in the previous section are covered by the 15 papers forming the second part of this thesis. The contributions of these papers are summarized in this section. The papers for each research area can be divided into two groups: one group proposes *theoretical findings* and *algorithms* derived on the basis of these findings, while the other group investigates a dedicated practical *application*. In Figure 1.3, the papers contained in each of the three research areas as well as the dependencies between the research areas are depicted.

1.3.1 Gaussian Filtering

Gaussian Filtering using State Decomposition Methods

In Paper A,

> F. Beutler, M. F. Huber, and U. D. Hanebeck. Gaussian Filtering using State Decomposition Methods. In *Proceedings of the 12th International Conference on Information Fusion (FUSION)*, pages 579–586, Seattle, WA, USA, July 2009,

LRKFs are extended in such a way that the number of samples used can be reduced significantly by exploiting special structures in both the dynamic model and measurement model. For this purpose, the state vector

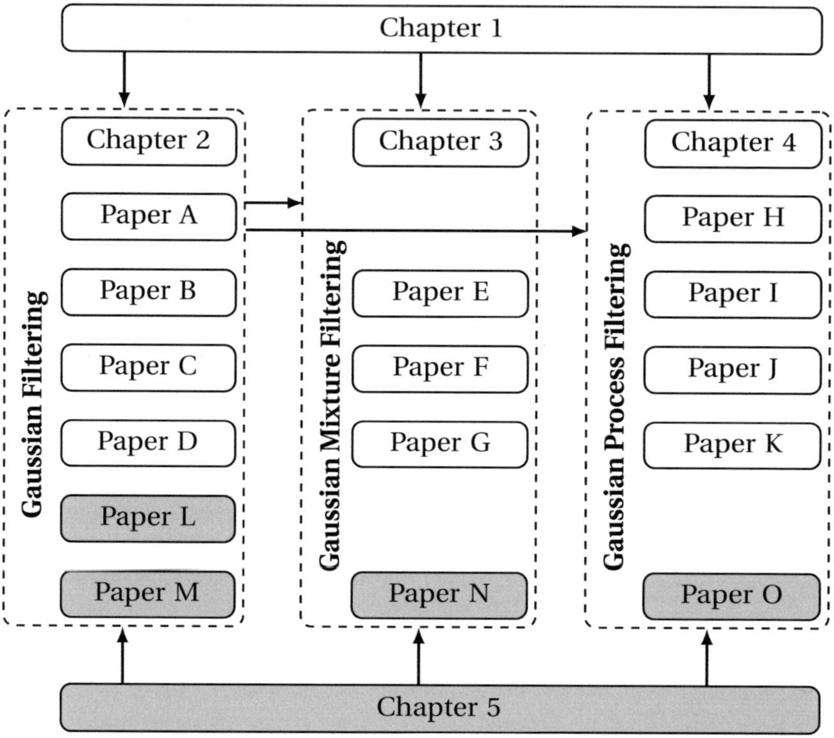

Figure 1.3: Structure of the thesis. Gray boxes indicate applications.

is decomposed in two ways: First, it is exploited that only some parts of the state vector are observable by measurements. Second, the models are decomposed in linear and nonlinear parts, where merely the nonlinear part is treated approximately by means of LRKFs. It shown by means of simulations and experiments that the estimation performance of the decomposed filters is comparable to the full-state ones, but the computation time is significantly reduced.

Semi-Analytic Gaussian Assumed Density Filter

The findings of Paper A are extended in Paper B,

> Marco F. Huber, Frederik Beutler, and Uwe D. Hanebeck. Semi-Analytic Gaussian Assumed Density Filter. In *Proceedings of the 2011 American Control Conference (ACC)*, pages 3006–3011, San Francisco, CA, USA, June 2011.

Instead of merely decomposing the models into linear and nonlinear substructures, a nonlinear-nonlinear decomposition is proposed, where one nonlinear part is conditionally integrable in closed form. This property holds only for special nonlinearities like polynomials, trigonometric functions, or squared exponential functions. For the conditionally integrable nonlinear part, mean vector and covariance matrix can be calculated analytically, if the filtering result is assumed to be Gaussian distribution. The other nonlinear part is still approximated via LRKFs. Simulations show an improved filtering accuracy and reduced computational demand.

Chebyshev Polynomial Kalman Filter

For polynomial nonlinearities closed-form moment propagation is possible, a property that was also exploited in Paper B. Paper C,

> M. F. Huber. Chebyshev Polynomial Kalman Filter. In *Digital Signal Processing*, vol. 23, no. 5, pages 1620–1629, September 2013,

leverages this property for arbitrary nonlinear systems. Here, these systems first are approximated by means of a Chebyshev polynomial series. The special structure of these orthogonal polynomials is then exploited for deriving very efficient closed-form vector-matrix expressions for mean and variance propagation. The superior performance of the resulting filter over the state-of-the-art is demonstrated via simulations and a real-world application.

Gaussian Filtering for Polynomial Systems Based on Moment Homotopy

All the algorithms proposed in Paper A–Paper C still rely on the assumption that state and measurement are joint Gaussian distributed. This assumption is relaxed in Paper D,

> M. F. Huber and U. D. Hanebeck. Gaussian Filtering for Polynomial Systems Based on Moment Homotopy. In *Proceedings to the 16th International Conference on Information Fusion (FUSION)*, pages 1080–1087, Istanbul, Turkey, July 2013,

for polynomial nonlinearities. This relaxation offers the opportunity of exactly determining the posterior mean and variance. However, a closed-form calculation is not possible and thus, a novel homotopy continuation method is proposed, which yields almost exact posterior mean and variance.

Application: Range-Based Localization

As application scenario for Gaussian filtering *range-based localization* is considered in Paper L,

> F. Beutler, M. F. Huber, and U. D. Hanebeck. Optimal Stochastic Linearization for Range-Based Localization. In *Proceedings of the 2010 IEEE/RSJ International Conference on Intelligent Robots and Systems (IROS)*, pages 5731–5736, Taipei, Taiwan, October 2010,

and in Paper M,

> F. Beutler, M. F. Huber, and U. D. Hanebeck. Semi-Analytic Stochastic Linearization for Range-Based Pose Tracking. In *Proceedings of the 2010 IEEE International Conference on Multisensor Fusion and Integration for Intelligent Systems (MFI)*, pages 44–49, Salt Lake City, UT, USA, September 2010.

In Paper L merely position and velocity of an object are estimated based on range measurements, which allows Gaussian assumed density filtering with closed-form calculation of mean and covariance. The orientation of an object is additionally considered in Paper M, which no longer allows an analytical solution. Instead, the nonlinear-nonlinear decomposition proposed in Paper B has to be employed.

1.3.2 Gaussian Mixture Filtering

(Semi-)Analytic Gaussian Mixture Filter

Closed-form mean and covariance calculation for special nonlinearities as well as the nonlinear-nonlinear model decomposition are extended in Paper E,

> M. F. Huber, F. Beutler, and U. D. Hanebeck. (Semi-)Analytic Gaussian Mixture Filter. In *Proceedings of the 18th IFAC World Congress*, pages 10014–10020, Milano, Italy, August 2011,

for Gaussian mixture filters. Both techniques can be employed component-wise and the superiority over state-of-the-art Gaussian mixture filters is demonstrated by means of simulations.

Adaptive Gaussian Mixture Filter Based on Statistical Linearization

Typically, Gaussian mixture filters based on the EKF or LRKFs assume a fixed number of mixture components, but determining the appropriate number requires to trade filtering performance off against computational load. In Paper F,

> M. F. Huber. Adaptive Gaussian Mixture Filter Based on Statistical Linearization. In *Proceedings of the 14th International Conference on Information Fusion (Fusion)*, Chicago, Illinois, July 2011,

an adaptive algorithm is proposed, which introduces new mixture components via splitting existing components. Splitting is performed whenever the nonlinearity of the dynamic or measurement model is high, but the current number of components is too low and thus, large (statistical) linearization errors are caused. Simulations show that new components are actually introduced where needed, while splitting at mild nonlinear or even linear parts of the models are avoided.

Superficial Gaussian Mixture Reduction

Due to adaptive splitting as in Paper F or due to the multiplication of Gaussian mixtures—which occurs if the transition density (1.6) or the likelihood (1.8) are also Gaussian mixtures—the number of mixture components grows unbounded. To limit this growth, in Paper G,

> M. F. Huber, P. Krauthausen, and U. D. Hanebeck. Superficial Gaussian Mixture Reduction. In *INFORMATIK 2011 - the 41th Annual Conference of the Gesellschaft für Informatik e.V. (GI), 6th Workshop Sensor Data Fusion: Trends, Solutions, Applications (SDF)*, Berlin, Germany, October 2011,

a reduction algorithm is proposed. It minimizes an upper bound of the curvature of the Gaussian mixture. This minimization can be formulated as a quadratic program that optimizes the component weights. The mixture is then reduced by removing component with zero weight. The advantages are an automated determination of the necessary number of components and a computational efficient implementation thanks to the plethora of solvers for quadratic programs.

Application: Gas Dispersion Source Estimation

The accurate and timely estimation of the location and strength of a gas release into atmosphere is imperative in order to increase the effectiveness of counter measures for protecting the public. In Paper N,

M. F. Huber. On-line Dispersion Source Estimation using Adaptive Gaussian Mixture Filter. In Proceedings of the *19th IFAC World Congress*, pages 1059–1066, Cape Town, South Africa, August 2014,

the adaptive Gaussian mixture filter proposed in Paper F is applied to this parameter estimation problem. In contrast to the Monte Carlo methods commonly used in this field, the proposed filter allows on-line estimation of the source parameters, while at the same time the estimation error is comparable or even lower than the state-of-the-art.

1.3.3 Gaussian Process Filtering

Analytic Moment-based Gaussian Process Filtering

Assuming that both the dynamic model and the measurement model are represented by means of Gaussian processes, Paper H,

M. P. Deisenroth, M. F. Huber, and U. D. Hanebeck. Analytic Moment-based Gaussian Process Filtering. In *Proceedings of the 26th International Conference on Machine Learning (ICML)*, Montreal, Canada, June 2009,

derives closed-form expressions for prediction and measurement update. Besides the joint Gaussian assumption of state and measurement, no additional approximations are employed.

Robust Filtering and Smoothing with Gaussian Processes

The results of Paper H are extended in Paper I,

M. P. Deisenroth, R. D. Turner, M. F. Huber, U. D. Hanebeck, and C. E. Rasmussen. Robust Filtering and Smoothing with Gaussian Processes. In *IEEE Transactions on Automatic Control*, vol. 57, no. 7, pages 1865–1871, July 2012, for smoothing. Assuming GPs with squared exponential covariance function and zero mean function, mean and covariance are derived analytically exactly and in closed form. It is further shown that restricting to this particular type of GP is not too restrictive as it corresponds to a universal function approximator.

Recursive Gaussian Process Regression

GPs are not well suited for applications, where data arrives during runtime. In Paper J,

> M. F. Huber. Recursive Gaussian Process Regression. In *Proceedings of the 38th International Conference on Acoustics, Sound, and Signal Processing (ICASSP)*, pages 3362–3366, Vancouver, BC, Canada, May 2013,

an on-line GP regression algorithm is proposed, which determines GP models recursively with a constant computation time. The key idea is to use a fixed-size set of so-called basis vectors that store all information necessary for regression. It is shown via synthetic and real-world data that this novel GP regression performs better than other on-line approaches. The resulting GP models can be utilized in the GP filtering algorithms proposed in Paper H and Paper I.

Recursive Gaussian Process: On-line Regression and Learning

In Paper J it is assumed that the hyperparameters of the GPs are known. This assumption is not valid in many applications. Alternatively, one has to learn the hyperparameters from data. Paper K,

> M. F. Huber. Recursive Gaussian Process: On-line Regression and Learning. *Pattern Recognition Letters*, vol. 45, pages 85–91, August 2014,

extends the recursive GP regression of Paper J with an on-line hyperparameter learning capability, i.e., regression and learning are performed simultaneously and during runtime. For this purpose, regression and learning are formulated as a Bayesian filtering problem, where the hyperparameters introduce nonlinearities. The nonlinear-nonlinear decomposition technique proposed in Paper B is employed to allow closed-form estimation of the hyperparameters.

Application: Bayesian Active Object Recognition via Gaussian Process Regression

In active object recognition, camera parameters like zoom or orientation are adapted to improve object classification performance. Due to the abstract nature of the mapping from object class—which corresponds to the system state—and image features, in Paper O,

> M. F. Huber, T. Dencker, M. Roschani, and J. Beyerer. Bayesian Active Object Recognition via Gaussian Process Regression. In *Proceedings of the 15th International Conference on Information Fusion (Fusion)*, pages 1718–1725, Singapore, July 2012,

this mapping is learned from data and represented by means of a GP. The classification task itself is formulated as a Bayesian filtering problem, for which closed-form expressions are derived. The appropriate camera parameters result from solving a sequential optimization problem that maximizes the mutual information between object class and image features.

1.4 Thesis Outline

This thesis consists of two parts: The first provides the background and the summaries of the aforementioned papers. The second part comprises the abstracts of the papers.

The structure of the thesis is depicted in Figure 1.3 on page 18. Chapter 2 briefly introduces the Gaussian density function and its properties. Also state-of-the-art Gaussian filters like the Kalman filter, LRKFs, or moment matching are described and summaries of the papers A–D are provided.

Gaussian mixture filtering is the content of Chapter 3. Besides the extension of Gaussian filters to Gaussian mixtures, this chapter also introduces the problem of increasing and reducing the number of mixture components. The chapter closes with summaries of the papers E–G.

In Chapter 4, Gaussian process regression and the state-of-the-art of Bayesian filtering with GP models is introduced. Furthermore, the complexity issues with GPs are discussed. This chapter comprises the summaries of the papers H–K.

Chapter 5 provides an introduction to each application treated with the algorithms developed in the papers A–K. It further summarizes the results of the papers L–O.

The first part of the thesis closes with Chapter 6, which gives a conclusion and an outlook to future work.

2

Gaussian Filtering

This chapter lays the foundation of the thesis. It first introduces the Gaussian distribution and some of its important properties in Section 2.1. Based on this, a general Gaussian filtering problem with focus on predictions and measurement updates is formulated in Section 2.2.1. This general problem can only be solved for some special cases, where the linear is the most prominent one (see Section 2.2.2). Otherwise, approximations have to be applied, where Sections 2.2.3–2.2.5 discuss the mostly utilized approximation techniques. In Section 2.3, the smoothing problem is formulated. The same approximation techniques introduced for prediction and measurement update can be applied here as well.

Many filtering problems, even though being nonlinear, comprise linear substructures. This special situation has been exploited in the past by means of the so-called Rao-Blackwellisation theorem. Section 2.4 provides a brief summary of this theorem and its application to Gaussian filtering. This theorem is also the starting point for the contribution made in this thesis for Gaussian filtering. The key findings of the Papers A–D are summarized in Section 2.5.

2.1 The Gaussian Distribution

The central distribution employed in this thesis is the Gaussian or normal distribution. It is also the most widely used distribution in statistics, filtering, and machine learning. Its probability density function is defined by

$$\mathcal{N}\left(\underline{x};\underline{\mu},\mathbf{C}\right) \triangleq \frac{1}{\sqrt{|2\pi\mathbf{C}|}} \mathrm{e}^{-\frac{1}{2}\left(\underline{x}-\underline{\mu}\right)^{\mathrm{T}}\mathbf{C}^{-1}\left(\underline{x}-\underline{\mu}\right)} \tag{2.1}$$

with the parameters mean vector and covariance matrix

$$\underline{\mu} = \mathrm{E}\left\{\underline{\boldsymbol{x}}\right\} = \int_{\mathbb{R}^{n_x}} \underline{x} \cdot \mathcal{N}\left(\underline{x};\underline{\mu},\mathbf{C}\right) \mathrm{d}\underline{x} ,$$
$$\mathbf{C} = \mathrm{Cov}\left\{\underline{\boldsymbol{x}}\right\} = \mathrm{E}\left\{\left(\underline{x}-\underline{\mu}\right)\left(\underline{x}-\underline{\mu}\right)^{\mathrm{T}}\right\} ,$$

respectively. The dimension of the mean vector corresponds to the dimension of $\underline{\boldsymbol{x}}$, which is n_x. The covariance matrix is symmetric and positive semi-definite, where the number of elements is the dimension of $\underline{\boldsymbol{x}}$ squared. The term $\sqrt{|2\pi\mathbf{C}|}$ in (2.1) is a normalization factor ensuring that the integral of the density is equal to one. If a random vector $\underline{\boldsymbol{x}}$ is Gaussian distributed, the term $\underline{\boldsymbol{x}} \sim \mathcal{N}\left(\underline{\mu},\mathbf{C}\right)$ expresses that $\underline{\boldsymbol{x}}$ is Gaussian distributed with parameters $\underline{\mu}$ and $\mathbf{C}$, where $\mathcal{N}\left(\underline{\mu},\mathbf{C}\right)$ is the (cumulative) Gaussian distribution function. If $\boldsymbol{x} \sim \mathcal{N}(0,1)$, $\boldsymbol{x}$ follows the *standard Gaussian* distribution.

In Figure 2.1, the density functions of various Gaussian random variables are depicted. It can be seen that the Gaussian density has only one single mode that is located at the mean $\underline{\mu}$.

2.1.1 Importance of the Gaussian

The Gaussian distribution is a universal tool in many fields. Besides its application in statistics and filtering as considered in this thesis, it is for example important in statistical mechanics to describe energy fluctuations while in biology it is used to describe population dynamics ([92],

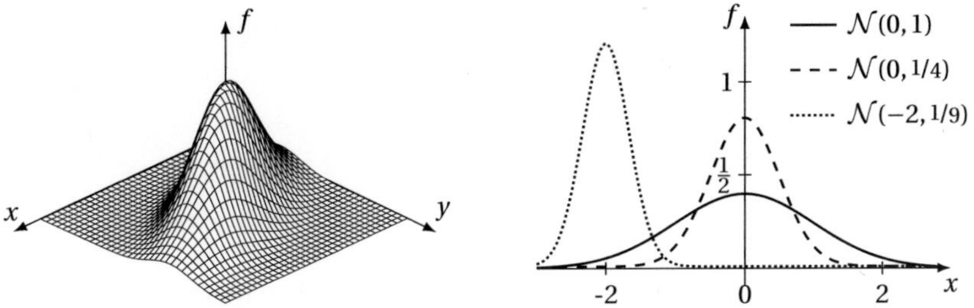

(a) Density function of a bivariate Gaussian.

(b) Density functions of different univariate Gaussians. The solid curve corresponds to the standard Gaussian.

Figure 2.1: Various univariate and bivariate Gaussian densities. Characteristic of Gaussian density functions is its *bell shape*.

Ch. 7). Also in economics it is the fundamental distribution, even though here the justification of applicability is sometimes questionable [191].

The success of the Gaussian distribution has many reasons, where the most important ones are the following [92, 101, 128]: First, it has merely two parameters and these are easy to interpret. These parameters at the same time correspond to the first two moments, which are the most basic properties of a distribution. Second, the number of parameters scales merely quadratically with the dimension of the state space. Third, if merely the first two moments are given, the Gaussian distribution makes the least assumptions about the true distribution according to the *maximum entropy principle* [19] and it is closest to the true distribution if the *Kullback-Leibler divergence* is considered as deviation measure [46, 74]. In parameter estimation it minimizes the *Fisher information*. Forth, it has a simple mathematical form, which allows closed-form calculations in many filtering problems and results in straightforward implementations. Fifth, according to the *central limit theorem* the sum of independent random variables approaches a Gaussian distribution, which makes the

Gaussian a preferred choice for modeling residual errors and noise. Sixth, Gaussians are unimodal and thus, they possess a single maximum. Such distributions are typical in many filtering problems like single-target tracking [197].

2.1.2 Dirac Delta Distribution

The Gaussian distribution possesses an important limiting case. If the determinant of the covariance matrix approaches zero, i.e., if $|\mathbf{C}| \rightarrow 0$, the Gaussian becomes a "peak" centered around the mean. This distribution is known as the *Dirac delta distribution* given by

$$\delta\big(\underline{x}-\underline{\mu}\big) = \begin{cases} \text{undefined} & \text{if } \underline{x}=\underline{\mu} \\ 0 & \text{if } \underline{x}\neq\underline{\mu} \end{cases}$$

such that

$$\int_{\mathbb{R}^{n_x}} \delta\big(\underline{x}\big)\,\mathrm{d}\underline{x} = 1\,.$$

The value of the Dirac delta is zero everywhere except of $\underline{x} = \underline{\mu}$. A useful property that follows from this fact is the so-called *sifting property*

$$\int_{\mathbb{R}^{n_x}} g\big(\underline{x}\big)\cdot\delta\big(\underline{x}-\underline{\mu}\big)\,\mathrm{d}\underline{x} = g\big(\underline{\mu}\big) \tag{2.2}$$

that selects only a single value from an integration.

2.1.3 The Exponential Family

The Gaussian density itself is a special case of two families of very general density functions: the Gaussian mixture densities and the exponential family. While the first family is treated in more detail in Chapter 3, a brief introduction to the second family is provided here.

A density function belongs to the *exponential family* if it is of the form

$$f(\underline{x}) = c(\underline{\eta}) \cdot h(\underline{x}) \cdot \exp\left(\underline{\eta}^{\mathrm{T}} \cdot \underline{\phi}(\underline{x})\right) , \tag{2.3}$$

where $\underline{\eta} \triangleq [\eta_0\ \eta_1\ \ldots\ \eta_n]^{\mathrm{T}}$ is a parameter vector, $\underline{\phi}(\underline{x})$ is a vector of *sufficient statistics*[1] comprising a set of n functions

$$\underline{\phi}(\underline{x}) \triangleq [\phi_0(\underline{x}) \quad \phi_1(\underline{x}) \quad \cdots \quad \phi_n(\underline{x})]^{\mathrm{T}} . \tag{2.4}$$

The term $c(\underline{\eta})$ is a normalization constant ensuring the probability mass being one and $h(\underline{x})$ is a scaling constant, often being one.

The exponential family comprises many well-known distributions as special cases. Examples are the Bernoulli distribution, Poisson distribution, multinomial distribution and of course the Gaussian distribution, which can be obtained from (2.3) by choosing

$$\underline{\eta} = \begin{bmatrix} \mathbf{C}^{-1} \cdot \underline{\mu} \\ -\frac{1}{2}\mathbf{C}^{-1} \end{bmatrix} , \quad \underline{\phi}(\underline{x}) = \begin{bmatrix} \underline{x} \\ \underline{x} \cdot \underline{x}^{\mathrm{T}} \end{bmatrix} , \quad c(\underline{\eta}) = \frac{1}{\sqrt{|2\pi\mathbf{C}|}} \mathrm{e}^{-\frac{1}{2}\underline{\mu}^{\mathrm{T}}\mathbf{C}^{-1}\underline{\mu}} , \quad h(\underline{x}) = 1 . \tag{2.5}$$

Furthermore, the exponential family is the only family of distributions with finite dimensional sufficient statistics and for which *conjugate priors* exist, i.e., the prior density has the same form as the likelihood, which simplifies Bayesian filtering significantly. However, except of some special cases like the Gaussian density, determining the moments of an exponential density in closed form is not possible in general [74].

1 A statistic $\phi(\mathcal{D})$ is said to be sufficient for data $\mathcal{D}$ if $f(x|\mathcal{D}) = f(x|\phi(\mathcal{D}))$, i.e., the statistic contains all information about the data such that the data can be discarded without loss of information.

2.2 Exact Gaussian Filtering and Approximations

In order to provide a solution of the general Bayesian filtering problem stated in Section 1.1, it is assumed in the following that the state vector $\underline{x}$ is Gaussian distributed. Thus, the filtering task boils down to calculating the two parameters mean vector and covariance matrix.

2.2.1 General Formulation

Various approaches have been proposed in the past for calculating the mean and covariance in case of arbitrary nonlinear dynamic and measurement models. To provide a unified overview of these approaches, the nonlinear transformation[2]

$$\underline{y} = \underline{g}(\underline{x}) + \underline{w}\,, \quad \underline{w} \sim \mathcal{N}(\underline{0}, \mathbf{C}_w)\,, \tag{2.6}$$

is considered in the following, where the Gaussian state $\underline{x} \sim \mathcal{N}(\underline{\mu}_x, \mathbf{C}_x)$ is mapped to a Gaussian random vector $\underline{y}$ via an arbitrary nonlinear function $\underline{g}(.)$. The noise $\underline{w}$ is independent of $\underline{x}$. For predictions, the transformation $\underline{g}(.)$ corresponds to the dynamic model $\underline{a}_k(.,.)$ in (1.4) for a given input $\underline{u}_k$ and $\underline{y}$, $\underline{w}$ are replaced by the predicted state $\underline{x}_{k+1}$ and the noise $\underline{w}_k$, respectively. In case of measurement updates, $\underline{g}(.)$ becomes the measurement model $\underline{h}_k(.)$, while $\underline{y}$ and $\underline{w}$ are replaced by the measurement vector $\underline{z}_k$ and noise $\underline{v}_k$, respectively.

For solving prediction and measurement update, the goal is to determine the joint Gaussian distribution of $\underline{x}$ and $\underline{y}$, which is given by

$$\begin{bmatrix} \underline{x} \\ \underline{y} \end{bmatrix} \sim \mathcal{N}(\underline{\mu}, \mathbf{C}) \quad \text{with } \underline{\mu} \triangleq \begin{bmatrix} \underline{\mu}_x \\ \underline{\mu}_y \end{bmatrix}, \mathbf{C} \triangleq \begin{bmatrix} \mathbf{C}_x & \mathbf{C}_{xy} \\ \mathbf{C}_{xy}^{\mathrm{T}} & \mathbf{C}_y \end{bmatrix}, \tag{2.7}$$

2 Only the additive noise case is discussed here. For non-additive noise, the solutions derived next can be directly applied if the state $\underline{x}$ is augmented with the noise $\underline{w}$, i.e., $\underline{g}(.)$ then becomes a function of both $\underline{x}$ and $\underline{w}$.

where

$$\begin{aligned}
\underline{\mu}_y &= \mathrm{E}\{\underline{g}(\underline{\boldsymbol{x}})\} = \int \underline{g}(\underline{x}) \cdot \mathcal{N}\left(\underline{x}; \underline{\mu}_x, \mathbf{C}_x\right) \mathrm{d}\underline{x} \,, \\
\mathbf{C}_y &= \mathrm{Cov}\{\underline{g}(\underline{\boldsymbol{x}})\} = \int \left(\underline{g}(\underline{x}) - \underline{\mu}_y\right)\left(\underline{g}(\underline{x}) - \underline{\mu}_y\right)^{\mathrm{T}} \cdot \mathcal{N}\left(\underline{x}; \underline{\mu}_x, \mathbf{C}_x\right) \mathrm{d}\underline{x} + \mathbf{C}_w \,, \\
\mathbf{C}_{xy} &= \mathrm{Cov}\{\underline{\boldsymbol{x}}, \underline{g}(\underline{\boldsymbol{x}})\} = \int \left(\underline{x} - \underline{\mu}_x\right)\left(\underline{g}(\underline{x}) - \underline{\mu}_y\right)^{\mathrm{T}} \cdot \mathcal{N}\left(\underline{x}; \underline{\mu}_x, \mathbf{C}_x\right) \mathrm{d}\underline{x} \,,
\end{aligned} \tag{2.8}$$

are the unknown mean vector and covariance matrix of $\underline{\boldsymbol{y}}$ as well as the unknown cross-covariance matrix between $\underline{\boldsymbol{x}}$ and $\underline{\boldsymbol{y}}$, respectively.

To perform a prediction, the respective parameters of the predicted Gaussian density can be retrieved from the first two lines of (2.8), namely the predicted mean and covariance. In case of a measurement update, an additional calculation is required as the posterior density is conditioned on the current measurement (see (1.7)). Thus, by conditioning $\underline{\boldsymbol{x}}$ on $\underline{\boldsymbol{y}}$ the resulting density is given by (see e.g. [144], Appendix A)

$$\underline{\boldsymbol{x}} | \underline{\boldsymbol{y}} \sim \mathcal{N}\left(\underline{\mu}_x + \mathbf{C}_{xy}\mathbf{C}_y^{-1} \cdot \left(\underline{y} - \underline{\mu}_y\right), \mathbf{C}_x - \mathbf{C}_{xy}\mathbf{C}_y^{-1}\mathbf{C}_{xy}^{\mathrm{T}}\right) , \tag{2.9}$$

which corresponds to desired *posterior density.*

Due to the restriction that both $\underline{\boldsymbol{x}}$ and $\underline{\boldsymbol{y}}$ are assumed to be Gaussian, a Bayesian filter calculating the *exact* joint Gaussian (2.7) is named a *Gaussian assumed density filter.* That is, although the true density is not Gaussian due to the nonlinear transformation (2.6), at least the first two moments—mean vector and covariance matrix—coincide with the true respective moments, why this approach often is also named *moment matching* [121]. Unfortunately, the integrals in (2.8) possess no analytic solution for arbitrary nonlinear functions $\underline{g}(.)$ in general and thus, approximations are inevitable. The various approximations proposed in the past for Gaussian filtering essentially differ in the way the integrals in (2.8) are solved. Just for some special cases, closed-form expressions for (2.8) can be found. The most prominent one is discussed next.

2.2.2 Linear Filtering

Assuming that the transformation in (2.6) is a linear one according to

$$\underline{y} = \mathbf{G} \cdot \underline{x} + \underline{w} ,$$

where the matrix $\mathbf{G}$ corresponds to the nonlinear function $g(.)$, the moment integrals in (2.8) can be calculated in closed form. The mean vector and covariance matrix of the joint Gaussian are then given by

$$\underline{\mu} = \begin{bmatrix} \underline{\mu}_x \\ \mathbf{G} \cdot \underline{\mu}_x \end{bmatrix} , \quad \mathbf{C} = \begin{bmatrix} \mathbf{C}_x & \mathbf{C}_x \mathbf{G}^{\mathrm{T}} \\ \mathbf{G}\mathbf{C}_x & \mathbf{G}\mathbf{C}_x \mathbf{G}^{\mathrm{T}} + \mathbf{C}_w \end{bmatrix} . \tag{2.10}$$

This follows directly from exploiting the linearity of the expectation operator E{.}. In doing so, the integrals in (2.8) are reduced to calculating the mean and covariance of $\underline{x}$.[3]

The Kalman Filter

Based on (2.10), the famous *Kalman filter* can be derived. For this purpose, linear dynamic and measurement models according to

$$\begin{aligned} \underline{x}_{k+1} &= \mathbf{A}_k \cdot \underline{x}_k + \underline{w}_k , \quad & \underline{w}_k &\sim \mathcal{N}\left(\underline{0}, \mathbf{C}_k^w\right) \\ \underline{z}_k &= \mathbf{H}_k \cdot \underline{x}_k + \underline{v}_k , \quad & \underline{v}_k &\sim \mathcal{N}\left(\underline{0}, \mathbf{C}_k^v\right) \end{aligned} \tag{2.11}$$

are assumed with system matrix $\mathbf{A}_k$ and measurement matrix $\mathbf{H}_k$. Given the posterior density $f_k^e(\underline{x}_k) = \mathcal{N}\left(\underline{x}_k; \underline{\mu}_k^e, \mathbf{C}_k^e\right)$ of the system state, the *prediction step* yields the predicted density $f_{k+1}^p(\underline{x}_{k+1}) = \mathcal{N}\left(\underline{x}_{k+1}; \underline{\mu}_{k+1}^p, \mathbf{C}_{k+1}^p\right)$ with parameters

$$\begin{aligned} \underline{\mu}_{k+1}^p &= \mathbf{A}_k \cdot \underline{\mu}_k^e , \\ \mathbf{C}_{k+1}^p &= \mathbf{A}_k \mathbf{C}_k^e \mathbf{A}_k^{\mathrm{T}} + \mathbf{C}_k^w , \end{aligned} \tag{2.12}$$

3 For calculating $\mathbf{C}_{xy}$ the identity $\mathrm{E}\{\underline{x} \cdot \underline{x}^{\mathrm{T}}\} = \mathbf{C}_x + \underline{\mu}_x \underline{\mu}_x^{\mathrm{T}}$ is required in addition.

where both the mean vector $\underline{\mu}^p_{k+1}$ and the covariance matrix $\mathbf{C}^p_{k+1}$ can be extracted from the second row in (2.10) by replacing $\underline{\mu}_x$, $\mathbf{C}_x$, $\mathbf{G}$, $\mathbf{C}_w$ with $\underline{\mu}^e_k$, $\mathbf{C}^e_k$, $\mathbf{A}_k$, $\mathbf{C}^w_k$, respectively.

For determining the *measurement update* of the Kalman filter, it is necessary to exploit the conditioning (2.9) by replacing $\underline{\mu}_x$, $\mathbf{C}_x$, $\mathbf{G}$, and $\mathbf{C}_w$ with $\underline{\mu}^p_k$, $\mathbf{C}^p_k$, $\mathbf{H}_k$, and $\mathbf{C}_v$, respectively. For a given measurement vector $\hat{\underline{z}}_k$, the resulting measurement update of the Kalman filter calculates the Gaussian posterior density $\mathcal{N}\left(\underline{x}_k; \underline{\mu}^e_k, \mathbf{C}^e_k\right)$ of the system state $\underline{\boldsymbol{x}}_k$ with mean vector and covariance matrix according to

$$
\begin{aligned}
\underline{\mu}^e_k &= \underline{\mu}^p_k + \mathbf{K}_k \cdot \left(\hat{\underline{z}}_k - \mathbf{H}_k \cdot \underline{\mu}^p_k\right) , \\
\mathbf{C}^e_k &= \mathbf{C}^p_k - \mathbf{K}_k \mathbf{H}_k \mathbf{C}^p_k ,
\end{aligned}
\tag{2.13}
$$

where $\mathbf{K}_k = \mathbf{C}^p_k \mathbf{H}^\mathrm{T}_k \left(\mathbf{H}_k \mathbf{C}^p_k \mathbf{H}^\mathrm{T}_k + \mathbf{C}^v_k\right)^{-1}$ is the so-called *Kalman gain*.

It is worth mentioning that R. Kalman used a different derivation in his paper [97]. Instead of the above approach, which is driven from a Bayesian perspective, he exploited orthogonal projections on the vector space spanned by the measurements. Both derivations are equivalent, where the Bayesian one makes the link to nonlinear Gaussian filters more obvious [160].

Route to Nonlinear Approaches

The additional assumption that not only $\underline{\boldsymbol{y}}$ but also the joint density of $\underline{\boldsymbol{x}}$ and $\underline{\boldsymbol{y}}$ is Gaussian as in (2.7) is only satisfied for linear models. However, all nonlinear Gaussian filtering approaches introduced next follow either implicitly or explicitly the same path that was utilized to derive the Kalman filter, i.e., constructing the joint Gaussian of $\underline{\boldsymbol{x}}$ and $\underline{\boldsymbol{y}}$ and conditioning on $\underline{\boldsymbol{y}}$. Thus, they all form a Kalman filter like approximation for nonlinear filtering problems. This approach is justified by the fact that by assuming both $\underline{\boldsymbol{x}}$ and $\underline{\boldsymbol{y}}$ being Gaussian, there *must* exist a linear transformation from $\underline{\boldsymbol{x}}$ to $\underline{\boldsymbol{y}}$ (see e.g. [198]). Hence, by providing an (approximate) solution to the integrals in (2.8), a linear transformation is constructed simultaneously that (implicitly) approximates the original nonlinear transformation (2.6). By explicitly calculating this linear transformation, the above Kalman filter equations can be applied directly. Hence, the focus in the next sections is mainly on how the approximation is achieved, while the actual filtering equations can be directly extracted from the above Kalman filter equations with appropriate substitution of the linear models (2.11).

Both approaches introduced next explicitly provide linear transformations for approximating the nonlinear filtering problem. They belong to the group of model approximating approaches depicted in Figure 1.2 on page 10. The filters in Section 2.2.5 instead can be considered as density approximating as they aim for the integrals in (2.8) in order to determine the joint Gaussian of $\underline{\boldsymbol{x}}$ and $\underline{\boldsymbol{y}}$. A linear model is merely calculated implicitly.

2.2.3 Linearized and Extended Kalman Filter

The motivation behind the linearized Kalman filter is that in case of mild nonlinearities it might be sufficient to linearize (2.6) about a nominal point through a Taylor-series expansion. Let $\underline{\bar{x}}$ be this nominal point. The Taylor-series expansion of the function $g(.)$ is then given by

$$\underline{\boldsymbol{y}} = \underline{g}(\underline{\boldsymbol{x}}) + \underline{\boldsymbol{w}} = \underline{g}(\underline{\bar{x}}) + \mathbf{G}_x(\underline{\bar{x}}) \cdot \Delta\underline{\boldsymbol{x}} + \mathcal{O}_x + \underline{\boldsymbol{w}} \,,$$

where $\Delta\underline{\boldsymbol{x}} \triangleq \underline{\boldsymbol{x}} - \underline{\bar{x}} \sim \mathcal{N}\left(\underline{\mu}_x - \underline{\bar{x}}, \mathbf{C}_x\right)$, $\mathbf{G}_x(\underline{\bar{x}})$ is the Jacobian matrix of $\underline{g}(.)$ according to

$$\mathbf{G}_x(\underline{\bar{x}}) \triangleq \left.\frac{\partial \underline{g}(\underline{x})}{\partial \underline{x}^{\mathrm{T}}}\right|_{\underline{x}=\underline{\bar{x}}} , \tag{2.14}$$

and $\mathcal{O}_x$ is the remainder comprising all higher-order terms of the expansion. A linear approximation of $\underline{g}(.)$ is obtained by neglecting the remainder term, which yields

$$\underline{g}(\underline{\boldsymbol{x}}) \approx \underline{g}(\underline{\bar{x}}) + \mathbf{G}_x(\underline{\bar{x}}) \cdot \Delta\underline{\boldsymbol{x}} \,. \tag{2.15}$$

Depending on the choice of the nominal point, different realizations of a linearized Kalman filter are obtained. In case of the basic linearized Kalman filter as described in [121] the nominal points are chosen in advance. In doing so, the linearized model (2.15) and thus, the Kalman gains and covariance matrices of the Kalman filter can be calculated off-line [210], which is desirable in applications with low computation resources.

A famous variation known as the *extended Kalman filter* (EKF)—the nonlinear filter used in the NASA Apollo program—is obtained when choosing the nominal points as being equal with the current state mean vector. This filter is no longer an off-line filter, but it is more adaptive to the current situation. This choice of nominal points also has implications on the moments of the joint Gaussian (2.7) as $\Delta\underline{\boldsymbol{x}} = \underline{\boldsymbol{x}} - \underline{\mu}_x \sim \mathcal{N}(\underline{0}, \mathbf{C}_x)$ now has zero mean. The mean vector $\underline{\mu}_y$ for instance is independent of the Jacobian $\mathbf{G}_x$ according to

$$\underline{\mu}_y = \mathrm{E}\left\{\underline{g}(\underline{\boldsymbol{x}}) + \underline{\boldsymbol{w}}\right\} \approx \mathrm{E}\left\{\underline{g}\left(\underline{\mu}_x\right) + \mathbf{G}_x\left(\underline{\mu}_x\right)\Delta\underline{\boldsymbol{x}} + \underline{\boldsymbol{w}}\right\} = \underline{g}\left(\underline{\mu}_x\right),$$

i.e., for calculating the mean it is sufficient to merely evaluate the nonlinear function $\underline{g}(.)$ at the nominal point $\underline{\mu}_x$.

A major advantage of linearized Kalman filters compared to Gaussian filters introduced below is their simplicity. Given the Jacobian matrix, they directly boil down to a Kalman filter. As a consequence, they are one of the computationally cheapest if not the cheapest Gaussian filters. However, differentiability of the nonlinear function $\underline{g}(.)$ is required, which is not given in every application.

Many improvements of linearized Kalman filters, but especially for the EKF have been suggested. The second-order EKF for instance utilizes a second-order Taylor-series expansion [67]. The iterated EKF performs multiple iterations of the measurement update for the same measurement value in order to linearize not only about the predicted state, but about the most recent estimate. This procedure converges faster to the exact solution than the EKF [93]. In [215] this concept has been generalized from discrete iterations to a differential update.

2.2.4 Statistical Linearization

An alternative to a Taylor-series expansion for explicitly determining a linear model is *statistical linearization* [67], where the nonlinear function $\underline{g}(.)$ is approximated via

$$\underline{g}(\underline{\boldsymbol{x}}) \approx \mathbf{G} \cdot \Delta\underline{\boldsymbol{x}} + \underline{b}$$

with $\Delta\underline{\boldsymbol{x}} \triangleq \underline{\boldsymbol{x}} - \underline{\mu}_x$. Here, the terms $\mathbf{G}$ and $\underline{b}$ are chosen in such a way that the mean squared error

$$\mathrm{E}\left\{\left(\underline{g}(\underline{\boldsymbol{x}}) - \mathbf{G} \cdot \Delta\underline{\boldsymbol{x}} - \underline{b}\right)^{\mathrm{T}} \left(\underline{g}(\underline{\boldsymbol{x}}) - \mathbf{G} \cdot \Delta\underline{\boldsymbol{x}} - \underline{b}\right)\right\} \tag{2.16}$$

is minimized with respect to $\mathbf{G}$ and $\underline{b}$. The solution to (2.16) yields

$$\underline{b} = \mathrm{E}\left\{\underline{g}(\underline{\boldsymbol{x}})\right\} , \tag{2.17}$$

$$\mathbf{G} = \mathrm{E}\left\{\underline{g}(\underline{\boldsymbol{x}}) \cdot \Delta\underline{\boldsymbol{x}}\right\} \mathbf{C}_x^{-1} . \tag{2.18}$$

There is an interesting relation between the linearized Kalman filter and statistical linearization. Assuming that $\underline{g}(.)$ is differentiable, the expectation in (2.18) can be reformulated to (see [160])

$$\mathrm{E}\left\{\underline{g}(\underline{\boldsymbol{x}}) \cdot \Delta\underline{\boldsymbol{x}}\right\} = \mathrm{E}\left\{\mathbf{G}_x(\underline{\boldsymbol{x}})\right\} \mathbf{C}_x ,$$

where $\mathbf{G}_x$ is the Jacobian matrix (2.14). In this case, the statistical linearization becomes a Taylor-series based linearization, except that instead of directly utilizing $\underline{g}(.)$ and the Jacobian $\mathbf{G}_x$, their expected values are employed. This is beneficial in the sense that statistical linearization exploits additional information about the state $\underline{\boldsymbol{x}}$ thanks to the expectation calculation, while for the Taylor-series expansion merely the mean vector $\underline{\mu}_x$ is utilized. This comes at the expense that the expected values (2.17) and (2.18) often cannot be calculated analytically, albeit the Jacobian matrix may exist.

2.2.5 Linear Regression Kalman Filters

To overcome the flaws of statistical and Taylor-series based linearization, the group of so-called *linear regression Kalman filters* (LRKFs) are based on a completely different approach. Instead of directly approximating the nonlinear model (2.6), the Gaussian representing the state $\underline{\boldsymbol{x}}$ is approximated by means of a set of weighted samples $\mathcal{L}_x = \{\omega_i, \mathcal{X}_i\}, i = 1 \ldots L$, which are sometimes also called *sigma points*. Given a sample representation of $\mathcal{N}(\underline{x}; \underline{\mu}_x, \mathbf{C}_x)$, the efficient evaluation of the integrals in (2.8) is straightforward.

In order to see this, the focus is restricted in the following on the integral

$$\mathrm{E}\left\{\underline{g}(\underline{\boldsymbol{x}})\right\} = \int \underline{g}(\underline{x}) \cdot \mathcal{N}\left(\underline{x}; \underline{\mu}_x, \mathbf{C}_x\right) \mathrm{d}\underline{x} \,. \tag{2.19}$$

The solution method for this integral can be directly applied to all integrals in (2.8). In order to obtain the sample representation independent of the current mean vector and covariance matrix of the Gaussian density, a change of the integration variable is employed for (2.19), which results in

$$\mathrm{E}\left\{\underline{g}(\underline{\boldsymbol{x}})\right\} = \int \underline{g}(\underline{x}) \cdot \mathcal{N}\left(\underline{x}; \underline{\mu}_x, \mathbf{C}_x\right) \mathrm{d}\underline{x} = \int \underline{g}\left(\underline{\mu}_x + \sqrt{\mathbf{C}_x} \cdot \underline{\theta}\right) \cdot \mathcal{N}\left(\underline{\theta}; \underline{0}, \mathbf{I}\right) \mathrm{d}\underline{\theta} \tag{2.20}$$

with $\sqrt{\mathbf{C}}$ being the matrix square root such that $\mathbf{C} = \sqrt{\mathbf{C}}\left(\sqrt{\mathbf{C}}\right)^{\mathrm{T}}$. This transformation allows determining the sample representation in advance for the multivariate standard Gaussian $\mathcal{N}\left(\underline{\theta};\underline{0},\mathbf{I}\right)$, while the adaptation to the actual Gaussian is performed on-line via the transformation $\underline{\mu}_x + \sqrt{\mathbf{C}_x} \cdot \underline{\theta}$.

With a given sample set $\mathcal{L}_\theta = \{\omega_i, \underline{\theta}_i\}, i = 1 \ldots L$, that approximates the standard Gaussian $\mathcal{N}\left(\underline{\theta};\underline{0},\mathbf{I}\right) \approx \sum_{i=1}^{L} \omega_i \cdot \delta\left(\underline{\theta} - \underline{\theta}_i\right)$ as a sum of weighted Dirac delta distributions, the integral (2.20) can be solved according to

$$\begin{aligned}
\mathrm{E}\left\{\underline{g}(\underline{\boldsymbol{x}})\right\} &= \int \underline{g}\left(\underline{\mu}_x + \sqrt{\mathbf{C}_x} \cdot \underline{\theta}\right) \cdot \mathcal{N}\left(\underline{\theta};\underline{0},\mathbf{I}\right) \mathrm{d}\underline{\theta} \\
&\approx \int \underline{g}\left(\underline{\mu}_x + \sqrt{\mathbf{C}_x} \cdot \underline{\theta}\right) \cdot \left(\sum_{i=1}^{L} \omega_i \cdot \delta\left(\underline{\theta} - \underline{\theta}_i\right)\right) \mathrm{d}\underline{\theta} \\
&= \sum_{i=1}^{L} \omega_i \cdot \underline{g}\underbrace{\left(\underline{\mu}_x + \sqrt{\mathbf{C}_x} \cdot \underline{\theta}_i\right)}_{\triangleq \mathcal{X}_i} ,
\end{aligned} \tag{2.21}$$

where the second line follows from substituting the standard Gaussian with its approximate sample representation. The third line follows from exploiting the sifting property (2.2).

The quantities $\mathcal{X}_i$ in (2.21) correspond to the *transformed sample points* with corresponding weights ω_i. The actual value of $\mathcal{X}_i$ depends on $\underline{\theta}_i$ and the way the matrix square root of $\mathbf{C}_x$ is determined. Also the weights ω_i offer an additional degree of freedom to the sample set. Various methods have been proposed in the past for calculating ω_i and $\underline{\theta}_i$. Main drivers for the calculation typically are numerical quadrature techniques for solving Gaussian integrals or capturing information of the standard Gaussian $\mathcal{N}\left(\underline{\theta};\underline{0},\mathbf{I}\right)$ like its mean and covariance.

Before the most popular of them are briefly introduced, it is worth mentioning that even though LRKFs resemble Monte Carlo integration in (2.21), the sample points are determined in a deterministic fashion. Monte Carlo methods like particle filters instead employ random sampling.

Cubature Kalman Filter (CKF)

The *cubature Kalman filter* [11, 214] exploits the third-order spherical cubature integration rule. According to this rule, the sample set consists of $L = 2 \cdot n_x$ points

$$\underline{\theta}_i = \begin{cases} \sqrt{n_x} \cdot \underline{e}_i & \text{if } i = 1 \ldots n_x \\ -\sqrt{n_x} \cdot \underline{e}_{i-n_x}, & \text{if } i = n_x + 1 \ldots 2n_x \end{cases} \tag{2.22}$$

with weights $\omega_i = 1/2n_x$ for all i. In (2.22), $\underline{e}_i = [0\,0\,1\,0\,0 \cdots 0]^{\mathrm{T}}$ is the canonical unit vector, where only element i is one. The CKF calculates (2.19) exactly if $\underline{g}(.)$ is a linear combination of monomials of order up to three [10], while the covariance $\mathbf{C}_x$ in (2.8) is determined exactly if $\underline{g}(.)$ is linear.

Unscented Kalman Filter (UKF)

As shown in [160], the CKF approach can be generalized to the so-called *unscented transform* proposed by [95], which forms the basis for the unscented Kalman filter [206]. Therefore, instead of $2n_x$ sample points, $2n_x + 1$ points are considered. The additional point is specially dedicated to the mean of $\underline{\boldsymbol{x}}$. The sample points and the corresponding weights are given by

$$\underline{\theta}_i = \begin{cases} \sqrt{n_x + \kappa} \cdot \underline{e}_i & \text{if } i = 1 \ldots n_x \\ -\sqrt{n_x + \kappa} \cdot \underline{e}_{i-n_x} & \text{if } i = n_x + 1 \ldots 2n_x \\ \underline{0} & \text{if } i = 2n_x + 1 \end{cases},$$

$$\omega_i = \begin{cases} \frac{1}{2(n_x+\kappa)} & \text{if } i = 1 \ldots n_x \\ \frac{1}{2(n_x+\kappa)} & \text{if } i = n_x + 1 \ldots 2n_x \\ \frac{\kappa}{n_x+\kappa} & \text{if } i = 2n_x + 1 \end{cases},$$

with κ being a free parameter. The placement of these sample points is depicted in Figure 2.2a on the next page. For $\kappa = 0$ the sample set is identical with the one provided by the CKF.

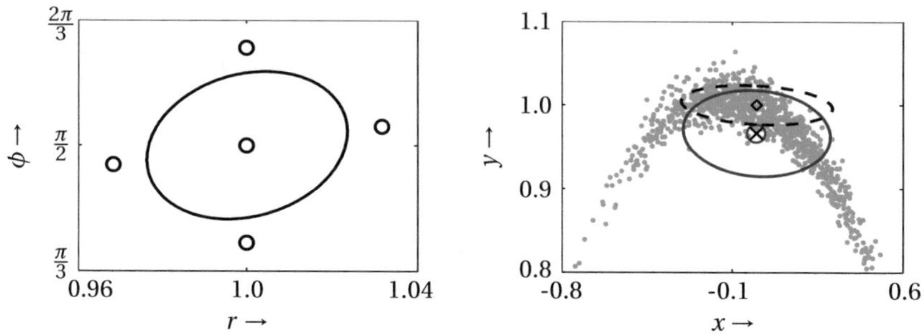

(a) Sample points of the UKF and covariance ellipse corresponding to the polar coordinates.

(b) True mean (black cross) as well as estimated mean and covariance of EKF (diamond for mean, dashed ellipse for covariance) and UKF (circle and solid gray ellipse).

Figure 2.2: Comparison of estimates of EKF and UKF. The gray dots forming a "banana" shape in (b) correspond to a Monte Carlo estimate of the true density.

Like for the CKF, it can be shown that the above sample set exactly captures the mean and covariance of the multivariate standard Gaussian. Furthermore, (2.19) is evaluated exactly if $\underline{g}(.)$ is a polynomial of order up to three.

Example 1: UKF vs. EKF

To demonstrate the difference in estimation between the UKF and the EKF, the nonlinear function

$$\begin{bmatrix} \boldsymbol{x} \\ \boldsymbol{y} \end{bmatrix} = \boldsymbol{r} \cdot \begin{bmatrix} \cos(\boldsymbol{\phi}) \\ \sin(\boldsymbol{\phi}) \end{bmatrix}, \quad \text{with} \quad \begin{bmatrix} \boldsymbol{r} \\ \boldsymbol{\phi} \end{bmatrix} \sim \mathcal{N}\left(\begin{bmatrix} 1 \\ \frac{\pi}{2} \end{bmatrix}, \begin{bmatrix} 0.0004 & 0.001 \\ 0.001 & 0.0685 \end{bmatrix} \right)$$

is considered (see [96]), which transforms the polar coordinates (r,ϕ) to Cartesian coordinates (x,y). The true density of the Cartesian coordinates is clearly not Gaussian as can be seen in Figure 2.2b.

However, the UKF almost exactly estimates the correct mean, while the estimate of the EKF is biased and inconsistent, especially in y direction. Furthermore, the EKF strongly underestimates the variance of the y coordinate.

Gaussian Estimator

The number of sample points of the CKF or UKF are restricted by the dimension of the state, although there are approaches to extend the sample set (see e.g. [193]). To allow an arbitrary number of samples, [89] proposed a sampling scheme that incorporates additional information about the shape of the Gaussian density function, besides merely capturing mean and covariance. The sample points are therefore given by

$$\underline{\theta}_i = \begin{cases} \mu_i \cdot \underline{e}_i & \text{if } i = 1 \ldots m \\ \mu_{i-m} \cdot \underline{e}_{i-m} & \text{if } i = m+1 \ldots 2m \\ \vdots & \\ \mu_{i-n_x \cdot m} \cdot \underline{e}_{i-(n_x-1)\cdot m} & \text{if } i = (n_x - 1)\cdot m + 1 \ldots n_x \cdot m \end{cases} \tag{2.23}$$

with weights $\omega_i = 1/(n_x \cdot m)$. Hence, the set of samples consists of $L = n_x \cdot m$ points, where m is a free parameter allowing to increase or decrease the required number of sample points. For each state dimension, the same parameters μ_i, $i = 1 \ldots m$, in (2.23) are required, which are the roots of the set of nonlinear equations

$$\frac{1}{2}\left(1 + \operatorname{erf}\left(\frac{\mu_i}{\sqrt{2}}\right)\right) - \frac{2i-1}{2m} + \lambda \mu_i = 0 \,,$$
$$\sum_{j=1}^{m} \mu_j^2 - m = 0 \,,$$

where erf(.) is the Gaussian error function and λ is a Lagrangian multiplier. Finding the roots is not possible in closed form and thus, requires a numerical solution. The necessary computational overhead is uncritical, as the sample set is determined for the multivariate standard Gaussian and can be transformed on-line according to (2.21).

Gauss-Hermite Kalman Filter (GHKF)

As can be seen from (2.23), all sample points are placed along the coordinate axes, while no points are placed in the quadrants. The same observation holds for CKF and UKF. A more dense placement of sample points as a irregular grid results when applying the *Gauss-Hermite quadrature* rule. Accordingly, the sample points and corresponding weights are

$$\underline{\theta}_{\underline{i}} = \left[\theta_{i_1}\ \theta_{i_2}\ \cdots\ \theta_{i_{n_x}}\right]^{\mathrm{T}}\ , \quad \omega_{\underline{i}} = \prod_{j=1}^{n_x} \frac{m!}{\left(m \cdot H_{m-1}(\theta_{i_j})\right)^2}\ , \tag{2.24}$$

with $\underline{i} \triangleq (i_1 i_2 \ldots i_{n_x})$ being an index vector where each index i_j takes values $1 \ldots m$. The jth element θ_{i_j}, $j = 1 \ldots n_x$, of the sample point $\underline{\theta}_{\underline{i}}$ is one of the m roots of m-order Hermite polynomial

$$H_m(x) = x \cdot H_{m-1}(x) - (m-1) \cdot H_{m-2}(x)\ , \quad m = 2, 3, \ldots \tag{2.25}$$

whereas this recursion commences from $H_0(x) = 1$ and $H_1(x) = x$.

The LRKF employing the above set of sigma points is named *Gauss-Hermite Kalman Filter* and was proposed in [13, 91]. Like the Gaussian estimator, the number of sample points can be varied. However, while for the Gaussian estimator the number of samples still scales linearly with the dimension of the state, the number of samples of the GHKF scales exponentially due the grid placement of the samples. This comes with the advantage that a GHKF using an m-order Hermite polynomial is exact is for polynomials up to order $2m - 1$.

Implicit Linearization

Besides the discussed LRKFs, there exist further approaches like the central difference filter [162] or the divided difference filter [134], which are not discussed here in order to constrain the focus on the mostly used approaches. However, all LRKFs share the property that they directly target the integrations (2.8). In doing so, also a linear transformation approximating the nonlinear one in (2.6) is constructed, although implicitly. As has been shown in [113, 204], LRKFs actually perform *weighted statistical linear regression* by approximating $\underline{g}(.)$ according to

$$\underline{g}(\underline{\boldsymbol{x}}) \approx \mathbf{G} \cdot \underline{\boldsymbol{x}} + \underline{b} \,, \tag{2.26}$$

where $\mathbf{G}$ and $\underline{b}$ result from minimizing

$$\sum_{i=1}^{L} \omega_i \cdot \left(\underline{g}(\mathcal{X}_i) - \mathbf{G} \cdot \mathcal{X}_i - \underline{b}\right)^{\mathrm{T}} \left(\underline{g}(\mathcal{X}_i) - \mathbf{G} \cdot \mathcal{X}_i - \underline{b}\right) .$$

The desired quantities are then given by

$$\mathbf{G} = \mathbf{C}_{xy}^{\mathrm{T}} \mathbf{C}_x^{-1} \,, \quad \underline{b} = \underline{\mu}_y - \mathbf{G} \cdot \underline{\mu}_x \,,$$

where

$$\underline{\mu}_x = \sum_i \omega_i \cdot \mathcal{X}_i \,, \qquad \mathbf{C}_x = \sum_i \omega_i \cdot \left(\mathcal{X}_i - \underline{\mu}_x\right)\left(\mathcal{X}_i - \underline{\mu}_x\right)^{\mathrm{T}} ,$$

$$\underline{\mu}_y \approx \sum_i \omega_i \cdot \underline{g}(\mathcal{X}_i) \,, \qquad \mathbf{C}_{xy} \approx \sum_i \omega_i \cdot \left(\mathcal{X}_i - \underline{\mu}_x\right)\left(\underline{g}(\mathcal{X}_i) - \underline{\mu}_y\right)^{\mathrm{T}}$$

are the sample means and covariances determined by means of the set of sample points $\mathcal{L}_x = \{\omega_i, \mathcal{X}_i\}$.

The error of the linearization (2.26) is given by

$$\underline{\boldsymbol{e}} = \underline{g}(\underline{\boldsymbol{x}}) - \mathbf{G} \cdot \underline{\boldsymbol{x}} - \underline{b} \tag{2.27}$$

and depends on the nonlinear function $\underline{g}(.)$ as well as on the state $\underline{\boldsymbol{x}}$. Both are main sources for linearization errors, which become large for strong nonlinearities or when the covariance of the state and thus its spread is large. It was shown in [113] that for LRKFs the error (2.27) has zero mean and a covariance matrix

$$\mathbf{C}_e = \mathbf{C}_y - \mathbf{G}\mathbf{C}_x\mathbf{G}^{\mathrm{T}} . \tag{2.28}$$

The latter is a potential and easy to evaluate indicator of the linearization error. If $\mathbf{C}_e$ is a zero matrix, the density of the error $\underline{\boldsymbol{e}}$ corresponds to a Dirac delta distribution [137] and the transformation $\underline{g}(.)$ is affine with $\underline{g}(\underline{\boldsymbol{x}}) = \mathbf{G}\cdot\underline{\boldsymbol{x}} + \underline{b}$. Accordingly, LRKFs are exact for linear/affine transformations and thus, degenerate to a standard Kalman filter.

2.3 Gaussian Smoothing

So far, the focus was on performing predictions and measurement updates for Gaussian filters. Now, the missing smoothing step is derived. According to (1.9), smoothing is a backward recursion for incorporating not only the current but also future measurements in the state density. For Gaussian filters it is assumed that the smoothed density is Gaussian, i.e.,

$$f_k^s(\underline{x}_k) = f_k^x(\underline{x}_k | \hat{\underline{z}}_{0:K}, \underline{u}_{0:K-1}) \approx \mathcal{N}(\underline{x}_k; \underline{\mu}_k^s, \mathbf{C}_k^s) \tag{2.29}$$

with appropriate mean vector $\underline{\mu}_k^s$ and covariance matrix $\mathbf{C}_k^s$. The derivation of the smoothing recursion for calculating (2.29) is based on the so-called *Rauch-Tung-Striebel smoother* (RTSS, see [145]) for linear systems and follows loosely [55, 85].

2.3.1 General Formulation

It is assumed that both the smoothed Gaussian $\mathcal{N}(\underline{x}_{k+1}; \underline{\mu}_{k+1}^s, \mathbf{C}_{k+1}^s)$ and the posterior Gaussian $\mathcal{N}(\underline{x}_k; \underline{\mu}_k^e, \mathbf{C}_k^e)$ are already given. As smoothing is an off-line task that is usually performed after $K \geq 1$ prediction and measurement update steps, the assumption of a known smoothed Gaussian is valid since for the time step K, the smoothed density coincides with the posterior Gaussian, i.e., $f_K^e(\underline{x}_k) \equiv f_K^s(\underline{x}_k) = \mathcal{N}(\underline{x}_K; \underline{\mu}_K^s, \mathbf{C}_K^s)$ and thus knowing $\mathcal{N}(\underline{x}_{k+1}; \underline{\mu}_{k+1}^s, \mathbf{C}_{k+1}^s)$ follows by induction.

Similar to Section 2.2.1, the interest is in the joint Gaussian

$$\begin{bmatrix} \underline{\boldsymbol{x}}_k \\ \underline{\boldsymbol{x}}_{k+1} \end{bmatrix} \sim \mathcal{N}\left(\underline{\mu}_x, \mathbf{C}_x\right) \quad \text{with } \underline{\mu}_x \triangleq \begin{bmatrix} \underline{\mu}_k^s \\ \underline{\mu}_{k+1}^s \end{bmatrix}, \mathbf{C}_x \triangleq \begin{bmatrix} \mathbf{C}_k^s & \mathbf{C}_{k|k+1} \\ \mathbf{C}_{k|k+1}^{\mathrm{T}} & \mathbf{C}_{k+1}^s \end{bmatrix}, \tag{2.30}$$

where the desired smoothed mean vector and covariance matrix for time step k can be retrieved from the first row of (2.30). To obtain both quantities, the joint Gaussian is rewritten as

$$\mathcal{N}\left(\left[\underline{x}_k, \underline{x}_{k+1}\right]^{\mathrm{T}}; \underline{\mu}_x, \mathbf{C}_x\right) = \mathcal{N}(\underline{x}_k; \underline{\mu}, \mathbf{C}) \cdot \mathcal{N}(\underline{x}_{k+1}; \underline{\mu}_{k+1}^s \mathbf{C}_{k+1}^s), \tag{2.31}$$

which is the product of the known smoothed Gaussian and the conditional Gaussian $\mathcal{N}(\underline{x}_k; \underline{\mu}, \mathbf{C}) \triangleq f(\underline{x}_k | \underline{x}_{k+1}, \underline{\hat{z}}_{0:k})$, where the latter is independent of the future measurements $\underline{\hat{z}}_{k+1:K}$ due to conditioning on $\underline{\boldsymbol{x}}_{k+1}$. However, the conditional Gaussian is unknown, but it can be obtained from (2.9) when replacing $\underline{\boldsymbol{x}}$ with $\underline{\boldsymbol{x}}_k \sim \mathcal{N}(\underline{\mu}_k^e, \mathbf{C}_k^e)$, $\underline{\boldsymbol{y}}$ with $\underline{\boldsymbol{x}}_{k+1} \sim \mathcal{N}(\underline{\mu}_{k+1}^p, \mathbf{C}_{k+1}^p)$ and the nonlinear transformation $\underline{g}(.)$ with $\underline{a}_k(.)$. Hence, the unknown mean vector and covariance matrix in (2.31) are given by

$$\begin{aligned} \underline{\mu} &= \underline{\mu}_k^e + \mathbf{J}_k \cdot \left(\underline{x}_{k+1} - \underline{\mu}_{k+1}^p\right), \\ \mathbf{C} &= \mathbf{C}_k^e - \mathbf{J}_k \mathbf{C}_{k|k+1}^{\mathrm{T}}, \end{aligned} \tag{2.32}$$

respectively, with gain matrix $\mathbf{J}_k = \mathbf{C}_{k|k+1} \left(\mathbf{C}_{k+1}^p\right)^{-1}$ and cross-covariance matrix $\mathbf{C}_{k|k+1}$ according to

$$\begin{aligned} \mathbf{C}_{k|k+1} &= \mathrm{Cov}\{\underline{\boldsymbol{x}}_k, \underline{\boldsymbol{x}}_{k+1}\} \\ &= \int \left(\underline{x}_k - \underline{\mu}_k^e\right) \left(\underline{a}_k(\underline{x}_k, \underline{u}_k) - \underline{\mu}_{k+1}^p\right)^{\mathrm{T}} \cdot \mathcal{N}\left(\underline{x}_k; \underline{\mu}_k^e, \mathbf{C}_k^e\right) \mathrm{d}\underline{x}_k . \end{aligned} \tag{2.33}$$

Substituting the mean and covariance of (2.32) into (2.31) and solving the product[4] leads to the desired joint Gaussian of (2.30), where the smoothed mean and covariance at time step k are given by

$$\begin{aligned} \underline{\mu}_k^s &= \underline{\mu}_k^e + \mathbf{J}_k \cdot \left(\underline{\mu}_{k+1}^s - \underline{\mu}_{k+1}^p\right) , \\ \mathbf{C}_k^s &= \mathbf{C}_k^e + \mathbf{J}_k \left(\mathbf{C}_{k+1}^s - \mathbf{C}_{k+1}^p\right) \mathbf{J}_k^{\mathrm{T}} , \end{aligned} \tag{2.34}$$

respectively. It is obvious that all quantities in (2.34) except of the gain matrix $\mathbf{J}_k$ are known from predictions and measurement updates or are calculated in a previous smoothing step. The matrix $\mathbf{J}_k$ requires solving the integral in (2.33), which is not possible in closed form in general due to the nonlinear system function $\underline{a}_k(.)$.

2.3.2 Linear Case

In case of linear dynamic and measurement models as in (2.11), a closed-form expression of the integral (2.33) and thus of the smoothing recursion (2.34) can be found. The resulting RTS smoother or sometimes *Kalman smoother* coincides with (2.34), where the gain matrix is

$$\mathbf{J}_k = \mathbf{C}_k^e \mathbf{A}_k^{\mathrm{T}} \left(\mathbf{C}_{k+1}^p\right)^{-1} \tag{2.35}$$

and the predicted as well as the posterior parameters are

$$\begin{aligned} \underline{\mu}_{k+1}^p &= \mathbf{A}_k \cdot \underline{\mu}_k^e , & \underline{\mu}_k^e &= \underline{\mu}_k^p + \mathbf{K}_k \cdot \left(\underline{\hat{z}}_k - \mathbf{H}_k \cdot \underline{\mu}_k^p\right) , \\ \mathbf{C}_{k+1}^p &= \mathbf{A}_k \mathbf{C}_k^e \mathbf{A}_k^{\mathrm{T}} + \mathbf{C}_k^w , & \mathbf{C}_k^e &= \mathbf{C}_k^p - \mathbf{K}_k \mathbf{H}_k \mathbf{C}_k^p . \end{aligned}$$

The latter correspond to the Kalman filter predictions (2.12) and measurement updates (2.13), respectively.

4 A detailed solution of this product can be found in Paper J.

2.3.3 Nonlinear Case

Similar to the predictions and measurement updates in case of nonlinear models, it is sufficient for Gaussian smoothing to find a linear approximation of the nonlinear system function $\underline{a}_k(.)$ or to solve the integral (2.33). Hence, all the nonlinear Gaussian filters discussed in Sections 2.2.3–2.2.5 can be employed, as they explicitly or implicitly provide a linear approximation and solve (2.33), respectively. Accordingly, for any of the nonlinear Gaussian filters a corresponding nonlinear Gaussian smoother can be found that exploits the respective approximation technique of the filter for determining the required quantities in (2.34). See for instance [8] for the extended Kalman smoother, [12] for the cubature Kalman smoother, or [159, 172] for the unscented RTS smoother.

2.4 Rao-Blackwellization

The perspective of solving a given Bayesian filtering problem was so far very coarse. Either the models were identified as being linear and the exact solution can be found, or the models are nonlinear and one has to rely on approximations. Actually, besides these black and white decisions, there are many gray scale values in between.

Example 2: Linear Substructure

Consider the measurement model

$$\underline{z} = \underline{h}(\underline{x}_n) + \mathbf{H}(\underline{x}_n) \cdot \underline{x}_l + \underline{v} , \tag{2.36}$$

where the measurement function comprises nonlinear parts denoted by $\underline{h}(.)$ as well as linear substructures indicated by the matrix $\mathbf{H}(.)$. Accordingly, the state $\underline{x}$ consists of two *sub-states* according to

$$\underline{x} = \begin{bmatrix} \underline{x}_l \\ \underline{x}_n \end{bmatrix} , \tag{2.37}$$

where $\underline{x}_n$ comprises the nonlinear state variables, while $\underline{x}_l$ comprises all state variables with *conditionally linear* dynamics, i.e., when conditioning on $\underline{x}_n$, the model (2.36) becomes a linear model.

By exploiting the linear substructure in the above example, it is possible to solve some of the filtering equations analytically, while an approximation is merely required for the nonlinear parts. The estimation performance gain of this decomposed processing is stated by the following theorem.

Theorem 1 (Rao-Blackwell, [128], Ch. 24) *Let* $\underline{x}$ *and* $\underline{y}$ *be dependent variables, and* $\underline{g}(\underline{x}, \underline{y})$ *be some scalar function. Then*

$$\mathrm{var}_{\underline{x},\underline{y}}\left\{g\left(\underline{x},\underline{y}\right)\right\} \geq \mathrm{var}_{\underline{x}}\left\{\mathrm{E}_{\underline{y}}\left\{g\left(\underline{x},\underline{y}\right)\middle|\underline{x}\right\}\right\} . \qquad \square$$

According to this so-called *Rao-Blackwell theorem*, a Bayesian filter utilizing this decomposition will result in a lower variance than a filter without it. This idea has been employed in many Bayesian filtering approaches like in particle filters [9, 41, 166] or in LRKFs [126]. Even though these filters employ this theorem mainly in case of linear-nonlinear substructures as in (2.36), it is important to note that this theorem is not restricted to those.

2.5 Contributions

In this section, the main contributions of the Papers A–D are summarized. The first two contributions in Section 2.5.1 and Section 2.5.2 exploited the aforementioned Rao-Blackwellization for Gaussian filtering. For Sections 2.5.3–2.5.5, it is assumed that the nonlinear function $\underline{g}(.)$ is either given as a polynomial and can be approximated by a polynomial. In doing so, efficient moment calculation algorithms are proposed.

2.5.1 Combining Rao-Blackwellization with Observed-Unobserved Decomposition

Measurement models with the same structure as in Example 2 are considered. In addition it is assumed that the state not only comprises the nonlinear and conditionally linear variables, but also contains state variables $\underline{x}_u$ that are *not directly observable*, i.e., there is no functional relation between the measurement vector $\underline{z}$ and $\underline{x}_u$ through the measurement model (2.36). Hence, the (predicted) state vector has the form

$$\underline{x}^p = \begin{bmatrix} \underline{x}_o^p \\ \underline{x}_u^p \end{bmatrix} \sim \mathcal{N}\left(\begin{bmatrix} \underline{\mu}_o^p \\ \underline{\mu}_u^p \end{bmatrix}, \begin{bmatrix} \mathbf{C}_o^p & \mathbf{C}_{ou}^p \\ \mathbf{C}_{uo}^p & \mathbf{C}_u^p \end{bmatrix} \right) , \tag{2.38}$$

where the *observed* part comprises the nonlinear and conditionally linear variables

$$\underline{x}_o^p = \begin{bmatrix} \underline{x}_n^p \\ \underline{x}_l^p \end{bmatrix} \sim \mathcal{N}\left(\begin{bmatrix} \underline{\mu}_n^p \\ \underline{\mu}_l^p \end{bmatrix}, \begin{bmatrix} \mathbf{C}_n^p & \mathbf{C}_{nl}^p \\ \mathbf{C}_{ln}^p & \mathbf{C}_l^p \end{bmatrix} \right) . \tag{2.39}$$

Such kind of state compositions appear in many application, where one is exemplified next.

Example 3: Object Localization

Consider a filtering problem, where the pose—that is location and orientation—and the corresponding velocities of an object in 3D are of interest. Sensors typically used for such localization problems are inertial sensors like gyroscopes and absolute localization techniques based on for instance multilateration. In such a situation, the system state may comprise the object pose $\underline{x}_n = \left[x\ y\ z\ \alpha\ \beta\ \gamma\right]^{\mathrm{T}}$, the translational velocities $\underline{x}_u = \left[\dot{x}\ \dot{y}\ \dot{z}\right]^{\mathrm{T}}$, and the angular velocities $\underline{x}_l = \left[\dot{\alpha}\ \dot{\beta}\ \dot{\gamma}\right]^{\mathrm{T}}$ (see e.g. [18]). Hence, even though the state as a whole is propagated via a dynamics model reflecting the motion of the object, the translational velocities cannot be observed directly.

Although there is no functional relation through the measurement model, the unobserved state variables are still updated thanks to the correlation between observed and unobserved states, which is reflected by the cross-covariance $\mathbf{C}^{ou}$. However, due to the missing functional relation, it seems to be a waste of computations to utilizes the state as a whole for evaluating the integrals (2.8), especially when the unobserved state is of high dimension. To reduce the computational load, a two-step decomposition of the filtering problem is proposed:

1. Updating the directly observed state $\underline{\boldsymbol{x}}_o$ first with the current measurement. In order to improve the estimation performance, the nonlinear and conditionally linear structure in accordance with Rao-Blackwellization is exploited.

2. Given the updated observed state, update the unobserved state $\underline{\boldsymbol{x}}_u$ by exploiting the correlation between $\underline{\boldsymbol{x}}_o$ and $\underline{\boldsymbol{x}}_u$.

The information flow of this two-step processing is depicted in Figure 2.3, which corresponds to the decomposition

$$f^e(\underline{x}) = f(\underline{x}|\hat{\underline{z}}) = f(\underline{x}_u, \underline{x}_o|\hat{\underline{z}}) = f(\underline{x}_u|\underline{x}_o) \cdot \underbrace{f(\underline{x}_o|\hat{\underline{z}})}_{=f(\underline{x}_o^e)=f(\underline{x}_n^e,\underline{x}_l^e|\hat{\underline{z}})} \tag{2.40}$$

of the conditional Gaussian (2.9).

Update of Observed State

Updating the observed state corresponds to calculating the conditional Gaussian $\underline{\boldsymbol{x}}_o^e \sim \mathcal{N}(\underline{\mu}_o^e, \mathbf{C}_o^e)$ according to (2.9), which requires to solve the integrals in (2.8). The procedure for this is shown exemplary for the measurement mean vector $\underline{\mu}_z$. Given the measurement function in (2.36), the mean vector is given by

$$\underline{\mu}_z = \int \left(\underline{h}(\underline{x}_n) + \mathbf{H}(\underline{x}_n) \cdot \underline{x}_l\right) \cdot \underbrace{f(\underline{x}_n, \underline{x}_l)}_{=f(\underline{x}_l|\underline{x}_n)\cdot f(\underline{x}_n)} \,\mathrm{d}\underline{x}_o \,, \tag{2.41}$$

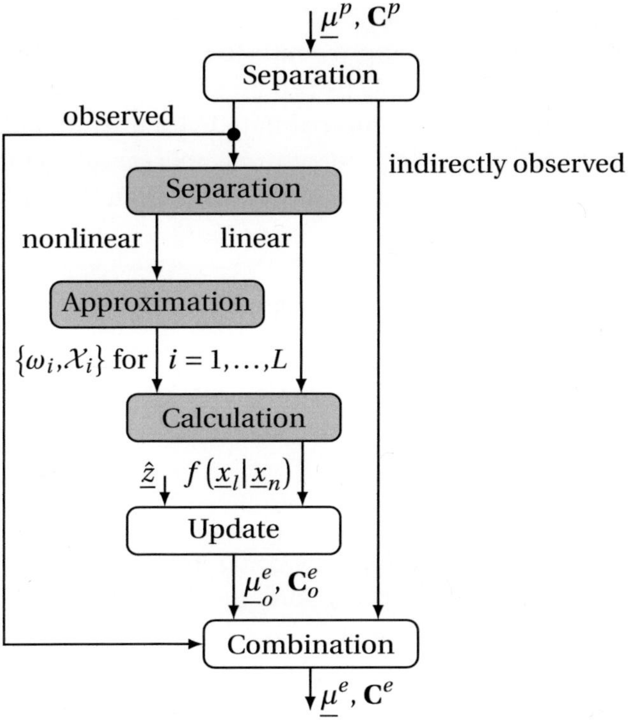

Figure 2.3: Information flow of the measurement update. The gray boxes indicate components that can be reused for the prediction step.

where $f(\underline{x}_l | \underline{x}_n) = \mathcal{N}\left(\underline{x}_l; \underline{\mu}_{l|n}(\underline{x}_n), \mathbf{C}_{l|n}\right)$ with mean vector and covariance matrix

$$\begin{aligned} \underline{\mu}_{l|n}(\underline{x}_n) &= \underline{\mu}_l^p + \mathbf{C}_{ln}^p \left(\mathbf{C}_n^p\right)^{-1} \left(\underline{x}_n - \underline{\mu}_n^p\right) , \\ \mathbf{C}_{l|n} &= \mathbf{C}_l^p - \mathbf{C}_{ln}^p \left(\mathbf{C}_n^p\right)^{-1} \mathbf{C}_{nl}^p , \end{aligned} \tag{2.42}$$

respectively. Hence, the mean of the conditional linear Gaussian $f(\underline{x}_l | \underline{x}_n)$ is a function of the nonlinear state.

Due to the nonlinear substate $\underline{\boldsymbol{x}}_n$, solving the above integral in closed form is not possible in general. To approximate the solution, the deterministic sampling techniques of the LRKFs can be employed here. In doing so, the Gaussian density $f(\underline{x}_n)$ is approximated by a mixture of Dirac delta distributions $\sum_i \omega_i \cdot \delta\left(\underline{x}_n - \mathcal{X}_i\right)$. Substituting this sample representation in (2.41) and utilizing the sifting properties of the Dirac delta distributions leads to

$$\begin{aligned}\underline{\mu}_z &= \sum_{i=1}^{L} \omega_i \cdot \int \left(\underline{h}(\mathcal{X}_i) + \mathbf{H}(\mathcal{X}_i) \cdot \underline{x}_l\right) \cdot f(\underline{x}_l | \mathcal{X}_i)\, \mathrm{d}\underline{x}_l \\ &= \sum_{i=1}^{L} \omega_i \cdot \underbrace{\left(\underline{h}(\mathcal{X}_i) + \mathbf{H}(\mathcal{X}_i) \cdot \underline{\mu}_{l|n}(\mathcal{X}_i)\right)}_{\triangleq \underline{\mu}_i^z} .\end{aligned}$$

The second equation follows from (2.42) and corresponds to a Kalman prediction thanks to the conditionally linear measurement function and the Gaussian density $f(\underline{x}_l | \mathcal{X}_i)$. Similarly the desired covariance $\mathbf{C}^z$ and cross-covariance $\mathbf{C}^{zx}$ are obtained

$$\begin{aligned}\mathbf{C}^z &= \sum_{i=1}^{L} \omega_i \cdot \left(\left(\underline{\mu}_i^z - \underline{\mu}_z\right)\left(\underline{\mu}_i^z - \underline{\mu}_z\right)^{\mathrm{T}} + \mathbf{H}(\mathcal{X}_i)\mathbf{C}_{l|n}\mathbf{H}(\mathcal{X}_i)^{\mathrm{T}}\right) , \\ \mathbf{C}^{oz} &= \sum_{i=1}^{L} \omega_i \cdot \left(\begin{bmatrix} \mathbf{O} \\ \mathbf{C}_{l|n}\mathbf{H}(\mathcal{X}_i)^{\mathrm{T}} \end{bmatrix} + \left(\begin{bmatrix} \mathcal{X}_i \\ \underline{\mu}_{l|n}(\mathcal{X}_i) \end{bmatrix} - \underline{\mu}_o\right)\left(\underline{\mu}_i^z - \underline{\mu}_z\right)\right) ,\end{aligned}$$

where again Kalman predictions have been used. Conditioning on the measurement $\underline{z}$ according to (2.9) yields the desired updated observed state $\underline{\boldsymbol{x}}_o^e$.

Update of Unobserved State

For updating the indirectly observed state, the mean vector $\underline{\mu}_o^e$ and covariance matrix $\mathbf{C}_o^e$ of the posterior Gaussian $f(\underline{x}_o | \hat{\underline{z}})$ are used. According to

[114], the mean vector of the unobserved state is updated via

$$\underline{\mu}_u^e = \underline{\mu}_u^p + \mathbf{J} \cdot \left(\underline{\mu}_o^e - \underline{\mu}_o^p\right) \tag{2.43}$$

with gain matrix $\mathbf{J} = \mathbf{C}_{uo}^p \left(\mathbf{C}_o^p\right)^{-1}$. The posterior covariance matrix of the unobserved state and cross-covariance matrix between observed and unobserved state are given by

$$\begin{aligned} \mathbf{C}_u^e &= \mathbf{C}_u^p + \mathbf{J}\left(\mathbf{C}_o^e - \mathbf{C}_o^p\right)\mathbf{J}^{\mathrm{T}}, \\ \mathbf{C}_{uo}^e &= \mathbf{J}\mathbf{C}_o^e, \end{aligned} \tag{2.44}$$

respectively. By comparison with (2.34), it is apparent that the above update equations coincide with an RTS smoother.

2.5.2 Semi-Analytical Filtering

The *semi-analytical Gaussian filter* (SAGF) introduced next takes Rao-Blackwellization to its extremes. Instead of merely restricting the decomposition to linear and nonlinear substructures, the SAGF exploits nonlinear-nonlinear decompositions, where for some nonlinear state variables an analytical solution of the Gaussian filtering problem can be found. Nonlinear functions for which analytical solutions exist are for example

- Monomials x^i with $i \in \mathbb{N}$,
- Trigonometric functions $\sin(x)$ and $\cos(x)$,
- Squared exponential functions $\exp\left(\underline{c}^{\mathrm{T}} \cdot \underline{\phi}(\underline{x})\right)$ with $\underline{c} \in \mathbb{R}^3$ and $\underline{\phi}(\underline{x}) \triangleq [1\ x\ x^2]^{\mathrm{T}}$,
- and linear combinations of the above functions,

assuming that the state density is Gaussian. The next example demonstrates the difference between a closed-form calculation of the moments in (2.8) for a quadratic function and the solution of an LRKF.

Example 4: Quadratic Transformation

For the quadratic transformation $\boldsymbol{y} = \boldsymbol{x}^2 + \boldsymbol{v}$ with $\boldsymbol{x} \sim \mathcal{N}(\mu_x, \sigma_x^2)$ and $\boldsymbol{v} \sim \mathcal{N}(0, \sigma_v^2)$ the moments in (2.8) can be calculated exactly to

$$\mu_y = \mu_x^2 + \sigma_x^2 \,, \quad \sigma_y^2 = 2\sigma_x^2 \cdot (\sigma_x^2 + 2\mu_x^2) + \sigma_v^2 \,, \quad \sigma_{xy} = 2 \cdot \sigma_x^2 \cdot \mu_x \,.$$

Using the UKF with parameter $\kappa = 0$ the same moments are calculated to

$$\mu_y = \mu_x^2 + \sigma_x^2 \,, \qquad \sigma_y^2 = 4\sigma_x^2 \cdot \mu_x^2 + \sigma_v^2 \,, \qquad \sigma_{xy} = 2 \cdot \sigma_x^2 \cdot \mu_x \,.$$

The mean μ_y is correct, which is not surprising as the unscented transform is exact for monomials up to order three. As the variance calculation for a quadratic function corresponds to an expectation calculation for a monomials of order four, the UKF introduces an error. More precisely, the variance σ_y^2 is underestimated as the term $2\sigma_x^4$ is missing and thus, the UKF is overconfident in this example.

On the one hand, analytical moment matching provides exact solutions to (2.8), but is restricted to a few nonlinear transformations, while on the other hand, LRKFs are generally applicable but may introduce severe linearization errors. The difference of LRKFs and analytical moment calculation from a linearization perspective is depicted in Figure 2.4.

The key idea of the SAGF is to combine both worlds by means of Rao-Blackwellization. Only some dimensions of the state $\underline{\boldsymbol{x}}$ are sampled via an LRKF and thus, only some parts of the nonlinear transformation (2.6)

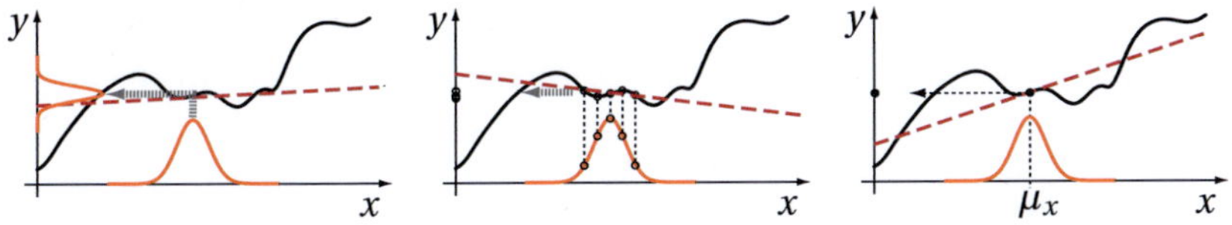

(a) Analytic Moment Matching **(b)** Sample-based linearization (LRKF) **(c)** Linearization via Taylor-series (EKF)

Figure 2.4: Illustration of different Gaussian filtering approaches: the nonlinear function (black) and its linearized versions (red dashed). Analytic moment matching utilizes the entire density $f(\underline{x})$ for (implicit) linearization, while the linearization of an LRKF is based on an approximate sample representation of $f(\underline{x})$. Thus, although the mean and covariance of $\underline{\boldsymbol{x}}$ are captured exactly by the samples, the same is not true for higher-order moment due to the finite number of samples. The EKF even linearizes the nonlinear function only around a single nominal point.

have to be treated approximately. For this purpose, the transformation is rearranged to the mapping

$$\underline{\boldsymbol{y}} = \underline{g}(\underline{\boldsymbol{x}}_a, \underline{\boldsymbol{x}}_s) + \underline{\boldsymbol{w}}\,, \tag{2.45}$$

where the Gaussian state $\underline{\boldsymbol{x}}^{\mathrm{T}} = [\underline{\boldsymbol{x}}_a^{\mathrm{T}}\ \underline{\boldsymbol{x}}_s^{\mathrm{T}}]$ consists of the substates $\underline{\boldsymbol{x}}_a$ (analytically integrable) and $\underline{\boldsymbol{x}}_s$ (sampled) with mean and covariance

$$\underline{\mu}_x = \begin{bmatrix} \underline{\mu}_a \\ \underline{\mu}_s \end{bmatrix}, \quad \mathbf{C}_x = \begin{bmatrix} \mathbf{C}_a & \mathbf{C}_{as} \\ \mathbf{C}_{sa} & \mathbf{C}_s \end{bmatrix}, \tag{2.46}$$

respectively. As there exists no closed-form expression for the desired moments (2.8), the decomposition into $\underline{\boldsymbol{x}}_a$ and $\underline{\boldsymbol{x}}_s$ is chosen in such a way that the moment integrals can be calculated analytically exactly for any given fixed value of $\underline{\boldsymbol{x}}_s$. Hence, the function $\underline{g}(.,.)$ is denoted to be *conditionally integrable*. For determining a sample-based representation of $\underline{\boldsymbol{x}}_s$, the sampling via LRKFs is applied.

It is worth mentioning that analytic moment matching and LRKFs are extreme cases of the SAGF: if $\underline{\boldsymbol{x}}_s$ is an empty vector, SAGF performs analytic moment matching and if $\underline{\boldsymbol{x}}_a$ is empty, the SAGF degenerates to an LRKF.

For the transformation given by (2.45), the moment calculation is shown exemplary for the mean vector $\underline{\mu}_y$

$$\underline{\mu}_y = \int \underline{g}(\underline{x}_a, \underline{x}_s) \cdot f(\underline{x}_a, \underline{x}_s) \, \mathrm{d}\underline{x} = \int \underline{g}(\underline{x}_a, \underline{x}_s) \cdot f(\underline{x}_a | \underline{x}_s) \cdot f(\underline{x}_s) \, \mathrm{d}\underline{x} \quad (2.47)$$

with the conditional Gaussian $f(\underline{x}_a | \underline{x}_s) = \mathcal{N}\left(\underline{x}_a; \underline{\mu}_{a|s}, \mathbf{C}_{a|s}\right)$ with mean and covariance

$$\begin{aligned} \underline{\mu}_{a|s} &= \underline{\mu}_a + \mathbf{C}_{as} \cdot \mathbf{C}_s^{-1} \cdot \left(\underline{x}_s - \underline{\mu}_s\right) , \\ \mathbf{C}_{a|s} &= \mathbf{C}_a - \mathbf{C}_{as} \cdot \mathbf{C}_s^{-1} \cdot \mathbf{C}_{sa} . \end{aligned} \quad (2.48)$$

By approximating the density $f(\underline{x}_s)$ of the sub-state $\underline{\boldsymbol{x}}_s$ with a mixture of Dirac delta distributions and by exploiting the sifting property, (2.47) simplifies to

$$\underline{\mu}_y \approx \sum_i \omega_i \cdot \underline{\mu}_i^y \quad \text{with} \quad \underline{\mu}_i^y = \int \underline{g}(\underline{x}_a, \mathcal{X}_i) \cdot f(\underline{x}_a | \mathcal{X}_i) \, \mathrm{d}\underline{x}_a .$$

It is important to note that this integral can be evaluated analytically as the function $\underline{g}(\cdot,\cdot)$ is conditionally integrable. Furthermore, solving these integrals is an off-line task and the solution is characterized by a parametric representation of the moments (2.48) and the sample points $\mathcal{X}_i$ for efficient on-line evaluation.

Example 5: Falling Body

To demonstrate improved estimation performance, the estimation of the altitude $\boldsymbol{\alpha}_k$, velocity $\boldsymbol{\beta}_k$, and constant ballistic coefficient $\boldsymbol{\gamma}_k$ of a

Table 2.1: Average rmse and its standard deviation over all simulation runs.

	Altitude	Velocity	Ballistic coefficient
SAGF	**12.6 ± 8.3**	**59.3 ± 143.7**	**0.016 ± 0.063**
UKF	14.2 ± 7.9	100.1 ± 212.4	**0.016 ± 0.058**
GPF 100 p.	13.0 ± 8.0	60.2 ± 134.9	0.029 ± 0.098
GPF 1000 p.	**12.7 ± 8.2**	**59.3 ± 142.1**	0.019 ± 0.066

falling body is considered [94, 174]. The system equation is given by

$$\underline{\boldsymbol{x}}_{k+1} = \begin{bmatrix} \boldsymbol{\alpha}_k \\ \boldsymbol{\beta}_k \\ \boldsymbol{\gamma}_k \end{bmatrix} + \Delta_t \cdot \begin{bmatrix} -\boldsymbol{\beta}_k \\ -\mathrm{e}^{-\rho\cdot\boldsymbol{\alpha}_k} \cdot (\boldsymbol{\beta}_k)^2 \cdot \boldsymbol{\gamma}_k \\ 0 \end{bmatrix} + \underline{\boldsymbol{w}}_k \,, \tag{2.49}$$

where $\underline{\boldsymbol{x}}_k = \left[\boldsymbol{\alpha}_k \; \boldsymbol{\beta}_k \; \boldsymbol{\gamma}_k\right]^{\mathrm{T}}$ is the state vector, $\Delta_t = 1$ the time discretization constant, $\rho = 5 \cdot 10^{-5}$ a constant factor. The noise $\underline{\boldsymbol{w}}_k$ is zero-mean Gaussian with covariance matrix $\mathbf{C}_k^w = 0.1 \cdot \mathbf{I}$. The initial state of the falling body is $\underline{x}_0^{\mathrm{T}} = [3 \cdot 10^5 \quad 2 \cdot 10^4 \quad 10^{-3}]$. The initial mean and covariance of the estimators for all simulation runs is set to be

$$\underline{\mu}^x = \begin{bmatrix} 3 \cdot 10^5 \\ 2 \cdot 10^4 \\ 10^{-5} \end{bmatrix} \,, \quad \mathbf{C}^x = \begin{bmatrix} 10^6 & 0 & 0 \\ 0 & 4 \cdot 10^6 & 0 \\ 0 & 0 & 20 \end{bmatrix} .$$

The state variables can be decomposed into $\underline{\boldsymbol{x}}_a = [\boldsymbol{\beta}_k \; \boldsymbol{\gamma}_k]^{\mathrm{T}}$ and $\boldsymbol{x}_s = \boldsymbol{\alpha}_k$. By conditioning on $\boldsymbol{x}_s$, the model (2.49) becomes a polynomial of order two.

A linear measurement equation is considered, where the altitude is measured directly according to

$$\boldsymbol{z}_k = \boldsymbol{\alpha}_k + \boldsymbol{v}_k \,, \quad \boldsymbol{v}_k \sim \mathcal{N}\left(0, \sigma_v^2\right) .$$

Due to the linearity of the measurement equation, the measurement update can be performed via the Kalman filter.

In Table 2.1, the average rmses of 1000 Monte Carlo simulation runs for three Gaussian filters, namely the proposed SAGF, the UKF, and the Gaussian particle filter (GPF, [106]), are listed. In case of the GPF 100 and 1000 particles are employed. The SAGF provides the most accurate estimate for all three state variables. Only the GPF with 1000 particles can compete with the SAGF, which however comes with a high computational load for the GPF. In terms of run time, the SAGF is two times faster than the GPF with 100 particles and four times faster than the UKF.

2.5.3 Chebyshev Polynomial Kalman Filtering

As mentioned in the previous section, analytic moment matching is for instance possible for polynomial nonlinearities. To apply this fact more generally, the following contribution consists of a two-step approach to allow Gaussian filtering for arbitrary nonlinear functions: First, the given nonlinear function is expanded in a series of Chebyshev polynomials. In the second step, which is discussed in more detailed in Section 2.5.4, exact expressions for the moment integrals (2.8) are provided in a computationally efficient vector-matrix notation. This approach named *Chebyshev polynomial Kalman filter* (CPKF) facilitates function approximation and Bayesian filtering in a black-box fashion without the need of manual operations or manual inspection similar to LRKFs, but with a potentially higher estimation performance.

Chebyshev Polynomials

The key idea behind the CPKF is the approximation of the nonlinear function (2.6) by means of a *truncated Chebyshev series expansion* according to

$$g(\boldsymbol{x}) \approx \sum_{i=0}^{n} c_i \cdot T_i(\boldsymbol{x}) \tag{2.50}$$

on the interval $\Omega \triangleq [-1,1]$, where $T_n(x)$ are Chebyshev polynomials of the first kind, which are defined compactly as

$$T_n(x) = \cos(n \cdot \arccos x)\ , \quad i = 0,1,\dots$$

or equivalently by means of the recursion

$$T_n(x) = 2x \cdot T_{n-1}(x) - T_{n-2}(x)\ , \quad n = 2,3,\dots\ , \tag{2.51}$$

with initial conditions

$$T_0(x) = 1\ , \quad T_1(x) = x\ . \tag{2.52}$$

It is easy to deduce from (2.51) that the function $T_n(x)$ is a polynomial of degree n. If n is even (odd), then $T_n(x)$ is a sum of even (odd) monomials, i.e., $T_n(x)$ is of the form

$$T_n(x) = \sum_{i=0}^{n} \alpha_{n,i} \cdot x^i = \sum_{j=0}^{\lfloor n/2 \rfloor} \alpha_{n,n-2j} \cdot x^{n-2j}\ , \tag{2.53}$$

where $\alpha_{n,i}$ is the *Chebyshev coefficient* of the ith monomial of the nth Chebyshev polynomial. The coefficient $\alpha_{n,i}$ is non-zero only if i is even (odd).

The quantities c_i in (2.50) are the *series coeffcients*

$$c_i = \frac{\langle g(x), T_i(x)\rangle}{\langle T_i(x), T_i(x)\rangle} \approx \tfrac{2-\delta_{0,n}}{n} \sum_{m=1}^{n} g(x_m) \cdot T_i(x_m) \tag{2.54}$$

for $i = 0,1,\dots,n$. In (2.54) $\langle g(x), f(x)\rangle \triangleq \int_\Omega \left(1-x^2\right)^{-1/2} \cdot g(x) \cdot f(x)\,\mathrm{d}x$. The right-hand side follows from thc discrctc orthogonality property of the Chebyshev polynomials, where $x_m = \cos(\pi(m-0.5)/i) \in \Omega$, $m = 1\dots i$, are the zeros of the Chebyshev polynomial $T_i(x)$ and $\delta_{i,j}$ is the Kronecker delta.

Employing polynomial series expansions in Bayesian filtering is not completely new. For instance higher-order Taylor-series expansion is employed for the second-order EKF [160, 174] or Fourier-Hermite series is used in Gaussian filtering in [161]. While the first approach is limited in terms of the order of the polynomial series, the second approach still requires numerical integration for moment calculation. Chebyshev series expansions are better suited for approximating nonlinear functions in the context of Gaussian filtering thanks to the following reasons:

- Chebyshev polynomials form a *complete orthogonal system* on Ω. As a consequence, the series coefficients (2.54) can be calculated independent of each other. This for instance is not true for Taylor-series expansions.
- In addition to the continuous orthogonality, Chebyshev polynomials are also *discrete orthogonal.* This property allows a very efficient calculation of the series coefficients by means of the well-known *discrete cosine transform* [30], for which a plethora of efficient algorithms exists. Similar approximate calculation schemes of series coefficients are typically not available for other orthogonal polynomials series like the Fourier-Hermite series.
- A truncated Chebyshev series satisfies the *near-minimax approximation* property, i.e., a truncated Chebyshev series of degree n is very close to the best possible polynomial approximation of the same degree. While the best polynomial representation of $g(.)$ is typically difficult to obtain, the Chebyshev series expansion is very close to the best solution and thanks to the discrete orthogonality very easy to calculate.

More detailed information about Chebyshev polynomials and their properties can be found for instance in [120].

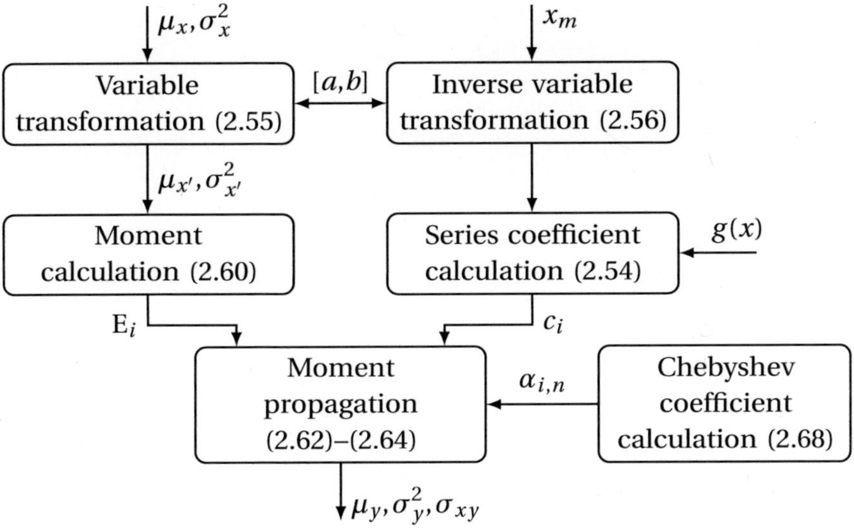

Figure 2.5: Flow chart of closed-form moment propagation for Chebyshev polynomial Kalman filter.

Structure

The building blocks of the CPKF are depicted in Figure 2.5. The blocks on the top are required due to the limitation that Chebyshev polynomials are only orthogonal on the interval Ω, while the function $g(.)$ can have an arbitrary support $[a,b] \subseteq \mathbb{R}$. Thus, the function $g(.)$ and the state $\boldsymbol{x}$ have to undergo first the affine transformation

$$x' = \tfrac{2}{b-a} \cdot x - \tfrac{a+b}{b-a} \,, \tag{2.55}$$

which yields a transformed Gaussian $\boldsymbol{x}' \sim \mathcal{N}\left(\mu_{x'}, \sigma_{x'}^2\right)$. Furthermore, the zeros x_m required for calculating the series coefficients (2.54) have to be mapped to the interval $[a,b]$, which is carried out by the inverse transformation

$$x = \tfrac{b-a}{2} \cdot x' + \tfrac{a+b}{2} \,. \tag{2.56}$$

of (2.55). By means of these transformations, arbitrary functions $g(.)$ can be treated by means of the CPKF.

The blocks "Moment calculation", "Moment propagation", and "Chebyshev coefficient calculation" are explained in detail in the following section.

2.5.4 Efficient Moment Propagation for Polynomials

At first a general polynomial representation of the function $g(.)$ according to

$$\boldsymbol{y} = g(\boldsymbol{x}) + \boldsymbol{w} = \sum_{i=0}^{n} c_i \cdot \boldsymbol{x}^i + \boldsymbol{w} \tag{2.57}$$

is considered. Chebyshev polynomials are treated as a special case at the end of this section.

General Solution

When propagating the Gaussian state $\boldsymbol{x}$ through the polynomial transformation $g(.)$ in (2.57), the mean of $\boldsymbol{y}$ can be expressed as

$$\mu_y = \mathrm{E}\{g(\boldsymbol{x})\} = \sum_{i=0}^{n} c_i \cdot \int x^i \cdot \mathcal{N}(x; \mu_x, \sigma_x^2)\, \mathrm{d}x = \sum_{i=0}^{n} c_i \cdot \underbrace{\mathrm{E}\{\boldsymbol{x}^i\}}_{\triangleq \mathrm{E}_i} . \tag{2.58}$$

Thus, the mean μ_y results in a weighted sum of non-central moments $\mathrm{E}_i = \mathrm{E}\{\boldsymbol{x}^i\}$ of order $i = 0,1,\ldots,n$. Formulae for calculating these moments of a Gaussian random vector are well-know (see for instance [117]), but require the evaluation of binomial coefficients, powers of the mean value, and weighted scalar products of the coefficient vector $\underline{\eta}$. Algebraically and computationally less demanding moment calculations can be found, however, when considering a special member of the exponential family in Section 2.1.3. By choosing the sufficient statistic to consist only of monomials of order up to n, i.e.,

$$\underline{\phi}(x) = \begin{bmatrix} 1 & x & x^2 & \cdots & x^n \end{bmatrix}^{\mathrm{T}} , \tag{2.59}$$

the following recursion proposed in [28] can be exploited. Although the moments of this special exponential density cannot be expressed in closed form, the ith order moment follows the recursion

$$\mathrm{E}_i = -\sum_{j=1}^{n} \frac{j}{i+1} \eta_j \, \mathrm{E}_{i+j} \; .$$

Thus, if the n lower-order moments $\underline{\mathrm{E}}_{0:n-1}^{\mathrm{T}} \triangleq [\mathrm{E}_0, \ldots, \mathrm{E}_{n-1}]$ are given and the moments up to E_m, $m \geq n$ are of interest, solving the linear system of equations

$$\mathbf{Q}(\underline{\eta}) \cdot \underline{\mathrm{E}}_{0:n-1} = \mathbf{R}(\underline{\eta}) \cdot \underline{\mathrm{E}}_{n:m} \tag{2.60}$$

gives the desired higher-order moments $\underline{\mathrm{E}}_{n:m}^{\mathrm{T}} \triangleq [\mathrm{E}_n \; \ldots \; \mathrm{E}_m]$. Here, $\mathbf{Q}(\underline{\eta})$ is an rectangular matrix and $\mathbf{R}(\underline{\eta})$ is a lower triangular matrix with elements (see [74])

$$Q_{i,j} = \begin{cases} 1 & \text{if } i = j \\ \frac{j-1}{i} \eta_{j-i} & \text{if } i < j \\ 0 & \text{otherwise} \end{cases} , \quad R_{i,j} = \begin{cases} 1 & \text{if } i - j = n \\ \frac{i-j-n}{i} \eta_{n+j-i} & \text{if } 0 \leq i - j < n \\ 0 & \text{otherwise} \end{cases} , \tag{2.61}$$

respectively. Thus, the matrix $\mathbf{R}(\underline{\eta})$ is zero everywhere except of the main diagonal and the n diagonals below the main. Thanks to this special structure, the linear system of equations (2.60) can be efficiently solved by means of forward substitution.

The Gaussian density is a special case of (2.59) for $\underline{\phi}(x) = \begin{bmatrix} 1 \; x \; x^2 \end{bmatrix}^{\mathrm{T}}$ where the first two moments $\mathrm{E}_0 = 1$ and $\mathrm{E}_1 = \mu_x$ are known. Hence, by solving

(2.60), the mean calculation (2.58) becomes

$$\mu_y = \underline{c}_n^{\mathrm{T}} \cdot \underline{\mathrm{E}}_{0:n} = \underline{c}_n^{\mathrm{T}} \cdot \begin{bmatrix} \mathbf{I}_2 \\ \mathbf{L} \end{bmatrix} \cdot \underline{\mathrm{E}}_{0:1} \,, \tag{2.62}$$

with $\mathbf{L} \triangleq \left(\mathbf{R}(\underline{\eta})\right)^{-1} \mathbf{Q}(\underline{\eta})$, where the parameter vector $\underline{\eta}$ comprises the parameters of a (normalized) Gaussian density as defined in (2.5). Furthermore, $\underline{c}_n^{\mathrm{T}} \triangleq [c_0\ c_1\ \dots\ c_n]$ is the vector of polynomial coefficients.

In a similar fashion, the variance σ_y^2 of $\boldsymbol{y}$ and the covariance σ_{xy} between $\boldsymbol{x}$ and $\boldsymbol{y}$ can be determined. The variance becomes

$$\begin{aligned} \sigma_y^2 &= \left(\underline{c}_n * \underline{c}_n\right)^{\mathrm{T}} \cdot \underline{\mathrm{E}}_{0:2n} - \mu_y^2 + \sigma_w^2 \\ &= \left(\mathbf{T} \cdot \underline{c}_n\right)^{\mathrm{T}} \cdot \underline{\mathrm{E}}_{0:2n} - \mu_y^2 + \sigma_w^2 \,, \end{aligned} \tag{2.63}$$

where $*$ is the discrete convolution operator. The second equality indicates an efficient matrix-vector realization of the convolution by means of the matrix $\mathbf{T}$ with entries $t_{i,j} = t_{i+1,j+1} = c_{i-j}$ if $i \in [j, j+n]$ and $t_{i,j} = 0$ otherwise, where $i = 1,2,\dots,2n+1$ and $j = 1,2,\dots,n+1$. Hence, $\mathbf{T}$ is special type of matrix, namely a triangular *Toeplitz matrix* with only the mean diagonal and n diagonals below the main diagonal being non-zero and all elements on individual diagonals being equal.

The covariance can be simplified to

$$\sigma_{xy} = \underline{c}_n^{\mathrm{T}} \cdot \underline{\mathrm{E}}_{1:n+1} - \mu_x \cdot \mu_y \,, \tag{2.64}$$

where μ_y is already known from (2.62). The first summand in (2.64) is almost identical to the mean calculation in (2.58) except for the shift by one in the order of the involved moments.

Given all the required moments, a Gaussian filter for polynomial nonlinearities is complete and listed in Algorithm 1. It is important to note that the polynomial order of the system function $a_k(.)$ and the measurement function $h_k(.)$ need not to be the same. Accordingly, the coefficient vec-

---

Algorithm 1 Polynomial Kalman Filter (PKF)

▷ *Prediction*

Determine moment vector $\underline{\mathrm{E}}_{0:2n_p}$ of posterior $\boldsymbol{x}_{k-1}^{e}$ by solving (2.60)

Predicted mean: $\mu_k^p = \left(\underline{c}_{n_p}^{p}\right)^{\mathrm{T}} \cdot \underline{\mathrm{E}}_{0:n_p}$

Predicted variance: $\left(\sigma_k^p\right)^2 = \left(\mathbf{T} \cdot \underline{c}_{n_p}^{p}\right)^{\mathrm{T}} \cdot \underline{\mathrm{E}}_{0:2n_p} - \left(\mu_k^p\right)^2 + \left(\sigma_k^w\right)^2$

▷ *Measurement Update*

Determine moment vector $\underline{\mathrm{E}}_{0:2n_e}$ of predicted state $\boldsymbol{x}_k^p$ by solving (2.60)

Measurement mean: $\mu_k^z = \left(\underline{c}_{n_e}^{e}\right)^{\mathrm{T}} \cdot \underline{\mathrm{E}}_{0:n_e}$

Measurement variance: $\left(\sigma_k^z\right)^2 = \left(\mathbf{T} \cdot \underline{c}_{n_e}^{e}\right)^{\mathrm{T}} \cdot \underline{\mathrm{E}}_{0:2n_e} - \left(\mu_k^z\right)^2 + \left(\sigma_k^v\right)^2$

Covariance: $\sigma_k^{xz} = \left(\underline{c}_{n_e}^{e}\right)^{\mathrm{T}} \cdot \underline{\mathrm{E}}_{1:n_e+1} - \mu_k^p \cdot \mu_k^z$

Kalman gain: $K_k = \sigma_k^{xz} / \left(\sigma_k^z\right)^2$

Posterior mean: $\mu_k^e = \mu_k^p + K_k \cdot \left(\hat{z}_k - \mu_k^z\right)$

Posterior variance: $\left(\sigma_k^e\right)^2 = \left(\sigma_k^p\right)^2 - K_k \cdot \sigma_k^{xz}$

---

tors $\underline{c}_{n_p}^{p}$ corresponding to the system function and $\underline{c}_{n_e}^{e}$ corresponding to the measurement function are of different dimension.

Example 6: Chaotic Synchronization

The proposed polynomial Kalman filter (PKF) is evaluated for the polynomial system model

$$\boldsymbol{x}_{k+1} = T_4(\boldsymbol{x}_k) + \boldsymbol{w}_k \tag{2.65}$$

as used in [117], where $T_i(x)$ is the ith Chebyshev polynomial (2.51). It is known that models as in (2.65) generate chaotic sequences [152], which are of practical use in securing communication systems. The true initial state $\boldsymbol{x}_0$ at time step $k = 0$ is assumed to be Gaussian with mean $\mu_0^x = 0.3$ and variance $\left(\sigma_0^x\right)^2 = 0.25$.

Furthermore, a linear measurement model

$$\boldsymbol{z}_k = \boldsymbol{x}_k + \boldsymbol{v}_k \tag{2.66}$$

is employed, with measurement noise variance $(\sigma_v)^2 = 10^{-2} \cdot (\sigma_w)^2$ and system noise variance being $(\sigma_w)^2 = 10^{-2}$ (high noise) or $(\sigma_w)^2 = 10^{-3}$ (low noise). The PKF is compared against EKF, UKF, and a particle filter (PF) with systematic resampling [37] and 500 samples. The latter is the only non-Gaussian filter. For all filters, 50 Monte Carlo simulation runs with identical noise sequences are performed, where the estimates are calculated for 50 time steps.

In Table 2.2, the average rmse, nees, and runtime over all Monte Carlo runs are listed for all filters and for both noise cases. For high noise, the proposed PKF outperforms all Gaussian filters in terms of rmse and nees, i.e., its estimates are closest to the true system state (low rmse) and at the same time the estimates are not overly confident (low nees). Furthermore, the matrix-vector terms proposed for the PKF allow for a runtime being close to the EKF, which is known to be the fastest Gaussian filter.

For the low noise case, UKF performs best in terms of estimation error, but PKF is very close to it. PF occasionally suffers from particle depletion, i.e., most of the particles converge towards the same state, which coincides with an overconfident estimate and thus an exceedingly high nees value. Even significantly increasing the number of particles or using different resampling techniques yields no improvement.

Table 2.2: Average rmse, nees, and runtime for system model (2.65).

	$\sigma_w^2 = 10^{-2}$				$\sigma_w^2 = 10^{-3}$			
	EKF	UKF	PF	PKF	EKF	UKF	PF	PKF
rmse	0.410	0.336	**0.292**	0.316	0.148	**0.118**	0.268	**0.118**
nees	4.737	1.550	1.168	**1.041**	7.279	**1.110**	–	1.129
time	**0.016**	0.038	0.109	0.017	**0.017**	0.037	0.102	0.018

Special Case: Chebyshev Polynomials

For a truncated Chebyshev series expansion of an arbitrary nonlinear function $g(.)$ as in (2.50) the moments calculations (2.62)–(2.64) could be applied directly. Therefore, the Chebyshev series has to be transformed into the standard polynomial form as in (2.57), for which the so-called *Clenshaw algorithm* [45] can be used. This procedure, however, has severe drawbacks: Evaluating Chebyshev series in the standard form (2.57) requires significantly more algebraic operations as the sparse structure of the Chebyshev polynomials is no longer exploited. Furthermore, the polynomial coefficients in (2.57) can be large numbers as they are products of multiple Chebyshev polynomial coefficients, which by themself already can be significant. For instance, the leading coefficient of $T_i(x)$ is 2^{i-1}.

To avoid these issues, it is recommended to reformulate (2.62)–(2.64) by exploiting the recursive definition of the Chebyshev polynomials. The mean μ_y for instance can be expressed as

$$
\begin{aligned}
\mu_y = \mathrm{E}\{g(\boldsymbol{x})\} &= \int g(x)\cdot \mathcal{N}\left(x;\mu_x,\sigma_x^2\right)\mathrm{d}x \\
&\overset{(2.50),(2.53)}{\approx} \sum_{i=0}^{n} c_i \sum_{j=0}^{i} \alpha_{i,j} \int x^j \cdot \mathcal{N}\left(x;\mu_x,\sigma_x^2\right)\mathrm{d}x \\
&= \underline{c}_n^{\mathrm{T}} \cdot \mathbf{A}_n \cdot \underline{\mathrm{E}}_{0:n}\ .
\end{aligned} \tag{2.67}
$$

In contrast to (2.62), $\underline{c}_n \triangleq [c_0\ c_1\ \dots\ c_n]^{\mathrm{T}}$ now is the vector of series coefficients (2.54). Further, $\mathbf{A}_n$ is the $(n+1)\times(n+1)$ matrix of Chebyshev coefficients defined by

$$\mathbf{A}_n \triangleq \begin{bmatrix}\underline{\alpha}_{0,n} & \underline{\alpha}_{1,n} & \cdots & \underline{\alpha}_{n,n}\end{bmatrix}^{\mathrm{T}} .$$

Here, $\underline{\alpha}_{i,n} \triangleq [\alpha_{i,0}\ \alpha_{i,1}\ \dots\ \alpha_{i,n}]^{\mathrm{T}} \in \mathbb{N}^{n+1}$, $i = 0,1,\dots,n$ comprises all coefficients of the ith Chebyshev polynomial up to and including the nth monomial. It is calculated via the recursion

$$\underline{\alpha}_{i,n} = 2\cdot\Big[0\ \underline{\alpha}_{i-1,i-1}^{\mathrm{T}}\ \underbrace{0\ \dots\ 0}_{n-i\text{ times}}\Big]^{\mathrm{T}} + \Big[\underline{\alpha}_{i-2,i-2}^{\mathrm{T}}\ \underbrace{0\ \dots\ 0}_{n-i+2\text{ times}}\Big]^{\mathrm{T}} , \tag{2.68}$$

where the recursion commences from

$$\underline{\alpha}_{0,n} = \begin{bmatrix}1 & 0 & \dots & 0\end{bmatrix}^{\mathrm{T}} , \quad \underline{\alpha}_{1,n} = \begin{bmatrix}0 & 1 & 0 & \dots & 0\end{bmatrix}^{\mathrm{T}}$$

and exploits the definition of the Chebyshev polynomials (2.51). According to (2.53), the coefficients $\alpha_{i,j}$ are zero for $j > i$. Thus, $\mathbf{A}_n$ is a sparse lower triangular matrix, which significantly reduces the computations of the matrix-vector products in (2.67).

Analogously, the variance σ_y^2 becomes

$$\sigma_y^2 \approx \left(\underline{c}_n \otimes \underline{c}_n\right)^{\mathrm{T}} \cdot \mathbf{P}_{2n} \cdot \underline{\mathrm{E}}_{0:2n} - \mu_y^2 + \sigma_w^2 , \tag{2.69}$$

with $\otimes$ being the Kronecker product and $\mathbf{P}_{2n}$ being an $(n+1)^2 \times (2n+1)$ matrix comprising the coefficients resulting from all possible products $T_i(x)\cdot T_j(x)$, $i,j = 0,1,\dots,n$ of the Chebyshev series expansion of $g(x)$.

Finally the covariance between $\boldsymbol{x}$ and $\boldsymbol{y}$ is given by

$$\sigma_{xy} = \underline{c}_n^{\mathrm{T}} \cdot \mathbf{A}_n^* \cdot \underline{\mathrm{E}}_{0:n+1} - \mu_x \cdot \mu_y , \tag{2.70}$$

where the $(n+1) \times (n+2)$ matrix $\mathbf{A}_n^*$ is given by

$$\mathbf{A}_{k,n}^* \triangleq \frac{1}{2}\left(\begin{bmatrix}\underline{0} & (b-a)\cdot\mathbf{A}_n\end{bmatrix} + \begin{bmatrix}(a+b)\cdot\mathbf{A}_n & \underline{0}\end{bmatrix}\right) .$$

This matrix includes the mapping back to the interval $[a, b]$ by means of the inverse variable transform (2.56).

Example 7: TV Commercial Effectiveness

In this example, real-world data from monitoring the advertising effectiveness of a TV commercial campaign for a single product is considered [124, 209]. This data is obtained by means of weekly surveys, where a given number of individuals from the population of TV viewers in UK is sampled in order to count the number being aware of current or recent TV commercials for the product. The result of each survey is measured in standardized units known as television ratings (TVRs) denoted by u_k.

The TVR measurements drive the nonlinear dynamics equation

$$\underline{\boldsymbol{x}}_{k+1} = \underline{a}\left(\underline{\boldsymbol{x}}_k + \underline{\boldsymbol{w}}_k, u_k\right), \quad \underline{\boldsymbol{w}}_k \sim \mathcal{N}\left(\underline{0}, 0.03 \cdot \mathbf{C}_k^x\right),$$

with system function

$$\underline{a}(\underline{x}, u) = \begin{bmatrix}\boldsymbol{x}_1 & \boldsymbol{x}_2 & \boldsymbol{x}_3 & \boldsymbol{x}_4 \\ & & & (\boldsymbol{x}_2 - \boldsymbol{x}_1) - (\boldsymbol{x}_2 - \boldsymbol{x}_1 - \boldsymbol{x}_3 \cdot \boldsymbol{x}_5) \cdot \exp(-\boldsymbol{x}_4 \cdot u)\end{bmatrix}^{\mathrm{T}}.$$

The state vector $\underline{x} \in \mathbb{R}^5$ comprises the minimum level of awareness $\boldsymbol{x}_1$, maximum level of awareness $\boldsymbol{x}_2$, memory decay rate $\boldsymbol{x}_3$, penetration $\boldsymbol{x}_4$, and effect of TVR on the awareness $\boldsymbol{x}_5$ (for details see [209]). The measurement equation is given by

$$\boldsymbol{z}_k = \boldsymbol{x}_{1,k} + \boldsymbol{x}_{5,k} + \boldsymbol{v}_k = \mathbf{H} \cdot \underline{\boldsymbol{x}}_k + \boldsymbol{v}_k, \quad \boldsymbol{v}_k \sim \mathcal{N}(0, 0.05)$$

with $\mathbf{H} \triangleq [1, 0, 0, 0, 1]$, where z_k corresponds to the awareness proportion. The initial state estimate is given by $\underline{\boldsymbol{x}}_0 \sim \mathcal{N}\left(\underline{\mu}_0^x; \mathbf{C}_0^x\right)$ with mean vector and covariance matrix

$$\underline{\mu}_0^x = \begin{bmatrix} 0.10 \\ 0.85 \\ 0.90 \\ 0.02 \\ 0.30 \end{bmatrix} \quad \text{and } \mathbf{C}_0^x = \begin{bmatrix} 6.25 & 6.25 & 0 & 0 & 0 \\ 6.25 & 406.25 & 0 & 0 & 0 \\ 0 & 0 & 1 & 0 & 0 \\ 0 & 0 & 0 & 2.25 & 0 \\ 0 & 0 & 0 & 0 & 100 \end{bmatrix},$$

respectively.

As the system state has dimension five, the nonlinear-nonlinear decomposition proposed in Section 2.5.2 is employed with $\boldsymbol{x}_a \triangleq \boldsymbol{x}_4$ and $\underline{\boldsymbol{x}}_s^{\mathrm{T}} \triangleq [\boldsymbol{x}_1\ \boldsymbol{x}_2\ \boldsymbol{x}_3\ \boldsymbol{x}_5]$. The sampled state $\underline{\boldsymbol{x}}_s$ is processed by means of the UKF. Additionally a UKF, where the unscented transform is applied to all five state dimensions is used.

The predictions of the awareness proportion z_k of the CPKF and UKF are compared before updating the state estimates with the true awareness value $\hat{z}_k$. The true awareness proportion values for performing the update step are taken from [209]. It is important to note that for the weeks 42, 43, and 44 no awareness measurements are available.

In Figure 2.6, the true and predicted awareness proportions are depicted. It is obvious that the UKF behaves very unsteady and is heavily fluctuating. The resulting awareness predictions are very inaccurate. This effect can be explained by overly confident estimates, i.e., the covariance matrix of the system state contains too small variances.

The behavior of the CPKF is different, which is surprising as the CPKF is merely applied on $\exp(-\boldsymbol{x}_4 \cdot u)$, while the remaining parts of the system equation are processed via the UKF. Thus, the CPKF has a stabilizing effect on the UKF resulting in awareness predictions that

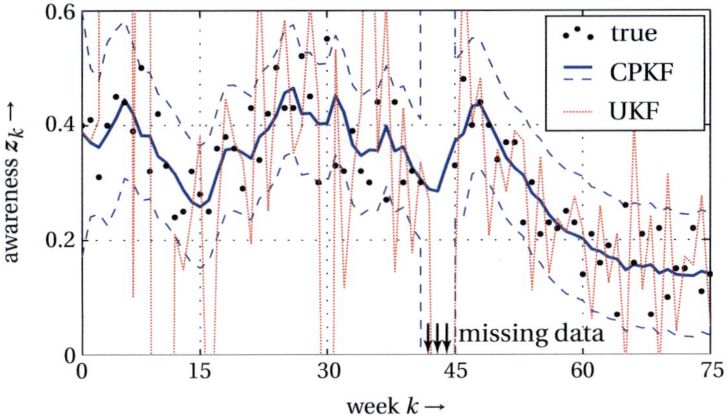

Figure 2.6: Predicted awareness proportions of the CPKF (blue solid line) with 95% confidence region (blue dashed) as well as the predictions of the UKF (red dotted). The true awareness proportion values are indicated by the black dots. For the weeks $k = 42, 43, 44$ no awareness measurements are available.

accurately follow the ground truth. Furthermore, the CPKF is not overconfident as the predicted measurement variances $\left(\sigma_k^z\right)^2$ are sufficiently large to capture the true awareness proportions. Even for the weeks with missing data, the predictions of CPKF are meaningful as the variances grow and thus, indicate an increasing uncertainty. Though, the trend is still correct.

2.5.5 Homotopic Moment Matching for Polynomial Measurement Models

Every Bayesian filter discussed so far in this chapter makes two different Gaussian assumptions. First, it assumes the predicted or posterior density to be Gaussian. Second, in order to perform the measurement update, it assumes that the joint density of state and measurement is Gaussian as well. The PKF for instance would be an exact Gaussian assumed density filter if only the first Gaussian assumption would be in place, as it performs moment matching, i.e., the mean and variance calculated by PKF coincide with the true mean and variance. The additional joint Gaussian assumption, however, can result in a poor approximation of the true mean and variance, which may cause a significant loss in estimation performance or even a divergence of the estimator.

Example 8: Joint Gaussian Flaw

To demonstrate the effect of the joint Gaussian assumption on the estimation performance, the polynomial measurement model

$$\boldsymbol{z} = \boldsymbol{x}^i + \boldsymbol{v} \tag{2.71}$$

is considered in the following, where $i > 0$ is even and the state is $\boldsymbol{x} \sim \mathcal{N}(0, \sigma_x^2)$. According to (2.62), (2.63), and (2.64), the mean μ_z, variance σ_z^2, and covariance σ_{xz} are given by

$$\mu_z = \mathrm{E}_i\ , \quad \sigma_z^2 = \mathrm{E}_{2i} - \mathrm{E}_i + \sigma_v^2\ , \quad \sigma_{xz} = \mathrm{E}_{i+1}\ , \tag{2.72}$$

respectively. Since $\boldsymbol{x}$ has zero mean, it follows that all even moments of $\boldsymbol{x}$ are non-zero and all odd moments are zero, i.e., $\mathrm{E}_i \neq 0$ and $\mathrm{E}_{i+1} = 0$ for all i being even. Hence, the covariance σ_{xz} in (2.72) is zero. As a result, the state $\boldsymbol{x}$ and measurement $\boldsymbol{z}$ are uncorrelated and the joint Gaussian of state and measurement is axis-aligned. In Figure 2.7a, the joint Gaussian for $i = 2$, $\sigma_x^2 = 1$, and $\sigma_v^2 = 0.1$ is depicted.

As the covariance σ_{xz} is zero, the Kalman gain K in line 8 of Algorithm 1 is zero as well and no update of the predicted state occurs, i.e., the posterior state $\boldsymbol{x}^e$ is identical to the predicted state $\boldsymbol{x}^p$. In this case, a given measurement value has no impact on the estimation.

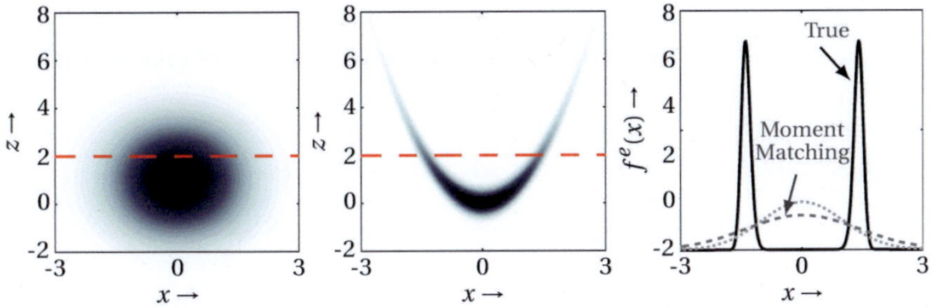

(a) Gaussian approximation of the joint density.

(b) True joint density.

(c) True posterior density (black) and Gaussian approximations.

Figure 2.7: Joint density $f(x,z)$ and posterior density $f^e(x)$ for $i=2$, i.e., for a quadratic polynomial. The red line indicates the measurement value $\hat{z}=2$. In (c), one Gaussian approximation is obtained based on the joint Gaussian assumption (dotted) and the other via moment matching (dashed), i.e., its mean and variance coincide with the true posterior moments.

Without the joint Gaussian assumption the missing update will not occur. In order to demonstrate this, the measurement update is now viewed from a full Bayesian perspective. Here, the posterior state $\boldsymbol{x}$ is represented by the posterior density $f^e(x)$ according to Bayes' rule (recall (1.7))

$$f^e(x) = \frac{f(z|x)\cdot f^p(x)}{f(z)} = \frac{f(x,z)}{f(z)}\,, \tag{2.73}$$

with the predicted Gaussian density $f^p(x) = \mathcal{N}\big(x; \mu^p, (\sigma^p)^2\big)$.

For the considered model (2.71) with a state $\boldsymbol{x}$ having zero mean, the joint Gaussian assumption leads to a factorization of the joint density $f(x,z) = f^p(x)\cdot f(z)$ as $\boldsymbol{x}$ and $\boldsymbol{z}$ are uncorrelated, which is equivalent to independence for Gaussian random variables. Hence, the Bayesian update in (2.73) degenerates to $f^e(x) = f^p(x)$. Actually, the joint density $f(x,z)$ is an exponential density with monomial sufficient statistics according to (2.59). This follows from the fact that the likelihood $f(z|x) = \mathcal{N}(z; x^i, \sigma_v^2)$ according to (1.8). The product of likelihood and prior density leads to the exponential density

$$\begin{aligned} f(x,z) &= f(z|x) \cdot f(x) = \mathcal{N}(z; x^i, \sigma_v^2) \cdot \mathcal{N}(x; 0, \sigma_x^2) \\ &= \exp\Big(-\log(2\pi\sigma_x\sigma_v) - \tfrac{1}{2\sigma_v^2} \cdot \big(z^2 + \tfrac{\sigma_v^2}{\sigma_x^2} x^2 - 2zx^i + x^{2i}\big)\Big) . \end{aligned} \tag{2.74}$$

This exponential joint density is depicted in Figure 2.7b for $i = 2$. By comparing Figure 2.7a with Figure 2.7b the difference between the true joint density and its Gaussian approximation becomes apparent. Given a measurement value $\hat{z} = 2$, Figure 2.7c depicts the posterior densities obtained for the Gaussian joint density and the true exponential joint density. It can be seen that the true posterior is bimodal, which only can be coarsely approximated by a Gaussian density. Furthermore, due to the joint Gaussian assumption, the Gaussian posterior does not even match the true posterior mean and variance.

Homotopy Continuation

To overcome the limitations of the joint Gaussian assumption, a new method for directly calculating the moments of the posterior density for *polynomial measurement models* $h(x) = \sum_{i=0}^{n_e} c_i^e \cdot x^i$ according to (2.57) will be introduced. This method does not require the joint Gaussian assumption and provides almost exact posterior mean and variance.

The key idea is to transform the known moments of the prior Gaussian density continuously into the desired posterior moments. For this purpose, *homotopy continuation* for calculating the moments of exponential densities as proposed in [147] is exploited. By means of a so-called progression parameter $\gamma \in [0\ 1]$ the posterior density $f^e(x)$ is parameterized in such a way that for $\gamma = 0$ the posterior density corresponds to prior Gaussian density $f^p(x)$ and for $\gamma = 1$ the posterior density corresponds to the true exponential density. For the initial value $\gamma = 0$, the moments are known as they coincide with the moments of the Gaussian prior. Incrementing the progression parameter causes moment variations described by means of a *system of ordinary differential equations* (ODEs). Solving this system of ODEs for $\gamma \in [0\ 1]$ gives the desired posterior moments.

To allow for homotopy continuation, the Bayesian measurement update is parametrized according to[5]

$$f^e(x;\gamma) \propto f\big(x,\hat{z};\underline{\eta}(\gamma)\big) \triangleq f(\hat{z}|x)^{\gamma} \cdot f^p(x) \tag{2.75}$$

for a given measurement value $\hat{z}$. Here, the *parametrized joint density* $f\big(x,\hat{z};\underline{\eta}(\gamma)\big) = \exp\big(\underline{\eta}(\gamma)^{\mathrm{T}} \cdot \underline{\phi}(x)\big)$ with $\underline{\phi}(x) \in \mathbb{R}^{2n_e+1}$ as in (2.59) is an exponential density similar to (2.74). The parameter vector is defined as

$$\underline{\eta}(\gamma) \triangleq \underline{\eta}^p + \gamma \cdot \underline{\eta}^l \in \mathbb{R}^{2n_e+1}$$

depending on γ. Here, $\underline{\eta}^p$ is the parameter vector of the Gaussian prior $f^p(x)$ and $\underline{\eta}^l$ is the parameter vector of the likelihood $f(\hat{z}|x)$.

In (2.75), $f^e(x;\gamma)$ is a parametrized version of the posterior density. For $\gamma = 1$, this parametrized measurement update corresponds to the standard Bayes' rule, while for $\gamma = 0$, the prior density $f^p(x)$ is directly assigned to the posterior density, i.e., no measurement update is performed.

System of Ordinary Differential Equations

By a continuous modification of the progression parameter γ, a continuous variation of the parameter vector $\underline{\eta}(\gamma)$ is achieved. This in turn results in a variation of the moments $\mathrm{E}_i\big(\underline{\eta}(\gamma)\big)$, $i = 0,\ldots,2n_e - 1$, of the parametrized joint density $f(x,\hat{z};\underline{\eta}(\gamma))$. These moment variations depending on γ can be described by means of a system of ODEs by calculating

5 To simplify the following calculations merely the proportional relation is considered. The normalization constant $1/\mathrm{E}_0 = 1/f(\hat{z})$ can be incorporated ex post without any disadvantages.

the partial derivatives $\dot{\mathrm{E}}_i \triangleq \partial \mathrm{E}_i(\underline{\eta}(\gamma))/\partial\gamma$ for $i = 0,\ldots,2n_e - 1$. The partial derivative of the ith-order moment is given by

$$\dot{\mathrm{E}}_i = \frac{\partial \mathrm{E}_i(\underline{\eta}(\gamma))}{\partial \gamma} = \begin{bmatrix} \mathrm{E}_i(\underline{\eta}(\gamma)) & \mathrm{E}_{i+1}(\underline{\eta}(\gamma)) & \cdots & \mathrm{E}_{i+2n_e}(\underline{\eta}(\gamma)) \end{bmatrix} \cdot \underline{\eta}^l \,, \tag{2.76}$$

which relates the variation of the ith-order moment to moments of order up to $i + 2n_e$. In the following, $\mathrm{E}_i^{(\gamma)} \triangleq \mathrm{E}_i(\underline{\eta}(\gamma))$ is used as shorthand term.

With the result in (2.76), the system of ODEs comprising the moment variations of all moments up to order $2n_e - 1$ is

$$\underline{\dot{\mathrm{E}}}_{0:2n_e-1} = \left(\mathbf{T}\left(\underline{\eta}^l\right)\right)^{\mathrm{T}} \cdot \underline{\mathrm{E}}^{(\gamma)}_{0:4n_e-1} = \mathbf{T}^l \cdot \underline{\mathrm{E}}^{(\gamma)}_{0:2n_e-1} + \mathbf{T}^h \cdot \underline{\mathrm{E}}^{(\gamma)}_{2n_e:4n_e-1} \,,$$

where $\mathbf{T}(\underline{\eta}^l) = \begin{bmatrix} \mathbf{T}^l \; \mathbf{T}^h \end{bmatrix}^{\mathrm{T}}$ is a Toeplitz matrix with entries $t_{i,j} = t_{i+1,j+1} = \eta^l_{i-j}$ if $i \in [j, j+2n_e]$ and $t_{i,j} = 0$ otherwise, where $i = 1,2,\ldots,4n_e$ and $j = 1,2,\ldots,2n_e$. Besides the lower-order moments $\underline{\mathrm{E}}^{(\gamma)}_{0:2n_e-1}$, the system of ODEs also depends on the higher-order moments $\underline{\mathrm{E}}^{(\gamma)}_{2n_e:4n_e-1}$. Fortunately, with the result of (2.60), the dependence on the higher-order moments can be resolved. In doing so, the system of ODEs can be reformulated into

$$\underline{\dot{\mathrm{E}}}_{0:2n_e-1} = \left(\mathbf{T}^l + \mathbf{T}^h \left(\mathbf{R}(\underline{\eta}(\gamma))\right)^{-1} \mathbf{Q}(\underline{\eta}(\gamma))\right) \cdot \underline{\mathrm{E}}^{(\gamma)}_{0:2n_e-1} \tag{2.77}$$

with starting solution $\underline{\mathrm{E}}^{(0)}{}_{0:2n_e-1}$ comprising the predicted moments. The matrices $\mathbf{R}(\underline{\eta}(\gamma))$ and $\mathbf{Q}(\underline{\eta}(\gamma))$ corresponding to (2.61), which vary with γ as they depend on the parameters of the parametrized joint density $f(x, \hat{z}; \underline{\eta}(\gamma))$.

The system of ODEs in (2.77) describes the moment variations caused by homotopy continuation of the Bayesian measurement update (2.75) in a very elegant manner. For solving this system of ODEs, standard numerical solvers based on the Runge-Kutta method [140] can be employed. The solution describes a trajectory of the moments $\underline{\mathrm{E}}^{(\gamma)}_{0:2n_e-1}$ depending on

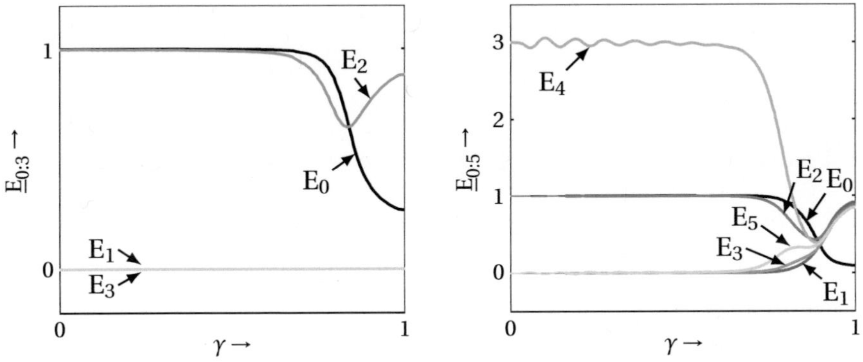

(a) Quadratic model $\boldsymbol{z} = \boldsymbol{x}^2 + \boldsymbol{v}$. (b) Cubic model $\boldsymbol{z} = \boldsymbol{x}^3 + \boldsymbol{v}$.

Figure 2.8: Trajectories of posterior moments.

different values of the progression parameter γ. The desired moments of the posterior density $f^e(x)$ are obtained for $\gamma = 1$, i.e., $\underline{\mathrm{E}}^{(1)}_{0:2n_e-1}$ comprises the result.

Example 9: Moment Trajectories

The polynomial measurement model of Example 8 is revisited, where now merely the quadratic (order $i = 2$) and the cubic (order $i = 3$) case are considered. Furthermore, the state $\boldsymbol{x} \sim \mathcal{N}(0,1)$ is standard Gaussian distributed, the measurement value is $\hat{z} = 1$, and $(\sigma_v)^2 = 0.1$ is the variance of the measurement noise. In Figure 2.8, the trajectories of the posterior moments resulting from the homotopy continuation are shown. It can be seen how the moments of the prior Gaussian are transformed into the true posterior moments.

As mentioned above, the moments in $\underline{\mathrm{E}}^{(1)}_{0:2n_e-1}$ are unnormalized as merely the proportional relation (2.75) was considered. Multiplying $\underline{\mathrm{E}}^{(1)}_{0:2n_e-1}$ with the normalization constant $\alpha \triangleq 1/f(\hat{z}) = 1/\mathrm{E}^{(1)}_0$ yields the actual posterior

Algorithm 2 Homotopic Polynomial Gaussian Filter (HPGF)

$\triangleright$ *Prediction*

1: Determine moment vector $\underline{\mathrm{E}}_{0:2n_p}$ of posterior $\boldsymbol{x}_{k-1}^e$ by solving (2.60)

2: Predicted mean: $\mu_k^p = \left(\underline{c}_{n_p}^p\right)^{\mathrm{T}} \cdot \underline{\mathrm{E}}_{0:n_p}$

3: Predicted variance: $\left(\sigma_k^p\right)^2 = \left(\mathbf{T} \cdot \underline{c}_{n_p}^p\right)^{\mathrm{T}} \cdot \underline{\mathrm{E}}_{0:2n_p} - \left(\mu_k^p\right)^2 + \left(\sigma_k^w\right)^2$

$\triangleright$ *Measurement Update*

4: Determine initial solution via first-order Taylor-series expansion around $\gamma = 0$

5: Solve system of ODEs (2.77) for $\gamma \in [\Delta\gamma; 1]$ with $\Delta\gamma \ll 1$

6: Calculate posterior mean $\mu_k^e = \alpha \cdot \mathrm{E}_1^{(1)}$

7: Calculate posterior variance $\left(\sigma_k^e\right)^2 = \alpha \cdot \mathrm{E}_2^{(1)} - \left(\mu_k^e\right)^2$

moments. The entire Gaussian filter employing moment homotopy is listed in Algorithm 2 and is named *homotopic polynomial Gaussian filter* (HPGF). The prediction step coincides with the prediction of the PKF since the PKF provides the exact predicted mean and variance. Before solving the ODE (2.77), a initialization step is required in line 4 as the matrix $\mathbf{R}\big(\underline{\eta}(\gamma)\big)$ is singular for $\gamma = 0$. Hence, the ODE is merely solved on the interval $\left[\Delta\gamma, 1\right]$, where $\Delta\gamma$ is a very small positive value. The posterior mean and variance calculated in line 6 and 7, respectively, are almost exact.

Example 10: Chaos Synchronization (Cont'd)

The estimation problem of Example 6 is revisited, but instead of the linear measurement model (2.66) the cubic measurement model

$$\boldsymbol{z}_k = \frac{\boldsymbol{x}_k^3}{20} + \boldsymbol{v}_k\,, \quad \text{with } \boldsymbol{v}_k \sim \mathcal{N}\left(0, 10^{-5}\right)$$

is employed. Given this measurement model, it turns out that *all* Gaussian filters relying on the joint Gaussian assumption diverge.

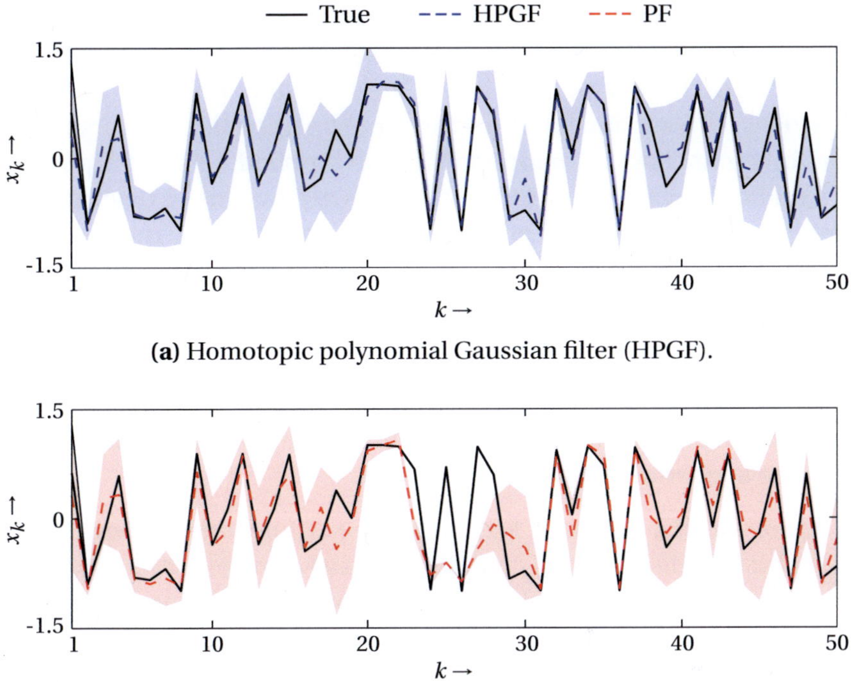

(a) Homotopic polynomial Gaussian filter (HPGF).

(b) Particle filter (PF).

Figure 2.9: State trajectory (black, solid line) and the estimates of HPGF and PF together with the corresponding 2-sigma confidence regions.

The HPGF, however, is able to provide valid estimates. In Figure 2.9a, an exemplary state trajectory is depicted. The estimates of HPGF accurately follow the true state. Furthermore, the true state is always within the 2-sigma confidence region of the estimates. The result of the PF depicted in Figure 2.9b is less accurate and shows sample depletion from time step $k = 20$ to $k = 27$.

2.6 Summary

In this chapter, the state-of-the-art in Gaussian filtering has been reviewed. Essentially, all Gaussian filters aim at approximating three particular moment integrals. Approximation techniques applied for this purpose are for instance linearization via Taylor-series expansions or deterministic sampling. The introduced approximation errors can be minimized by means of exploiting Rao-Blackwellization. This is starting point for the contributions made in this chapter:

- *Combining Rao-Blackwellization with decomposition of observed and unobserved states:* Unobserved state variables can be excluded from the approximation which reduces the computational load and the approximation error.
- *Rao-Blackwellization for nonlinear-nonlinear decomposition:* Instead of the commonly employed linear-nonlinear decomposition, a novel nonlinear-nonlinear decomposition is proposed. Therefore the concept of conditionally integrable functions was introduced, i.e., functions that comprise a nonlinear substructure for which analytic moment expressions exist.
- *Approximation of arbitrary nonlinear models with Chebyshev polynomials:* Polynomials are one class of functions for which analytic moment matching is possible. Chebyshev polynomials are well suited for transforming arbitrary nonlinearities into polynomials thanks to their orthogonality, sparseness, and efficient coefficient calculation procedure.
- *Closed-form moment calculation for polynomial nonlinearities:* Novel matrix-vector expressions for analytical moment calculation for polynomial dynamic and measurement models have been proposed. These expressions allow exact predictions for Gaussian filters.

- *Almost exact posterior moments by homotopy continuation:* For polynomial measurement models the joint Gaussian assumption can be avoided. For this purpose a homotopy continuation method was proposed that yields almost optimal posterior moments.

Every contribution on its own can significantly improve the estimation performance. A boosting in performance, however, can be obtained by exploiting the strong interrelations between the several methods. For instance, after employing the aforementioned decompositions there will remain nonlinear substructures for which no analytic moment matching is possible. However, approximating these nonlinearities with Chebyshev polynomials allows accurate Gaussian filtering for the entire nonlinear model as shown in Example 7. Given a polynomial measurement model—either polynomial by definition, generated via Chebyshev series expansion, or as a resulting substructure after Rao-Blackwellization—the novel moment homotopy can be employed.

3

Gaussian Mixture Filtering

Although Gaussian filters show a good estimation performance in many practical applications, they are clearly limited when the true distribution of the state takes a complex shape, e.g., multiple modes, strong skewness, or heavy tailes. Such situations may for instance arise in multi-target tracking or financial forecasts. To also provide a consistent filter in these applications, a natural extension of Gaussian filtering is Gaussian mixture filtering. In this chapter, a brief introduction to Gaussian mixture densities is given at first. Then the extension of the Gaussian filters discussed in Section 3.2 is derived. Due to the usage of multiple Gaussians, Gaussian mixtures filters require additional operations regarding the adaptation of the number of mixture components. A statement of this additional problem is given in the Section 3.3. The contributions made by the Papers E–G are summarized in Section 3.4.

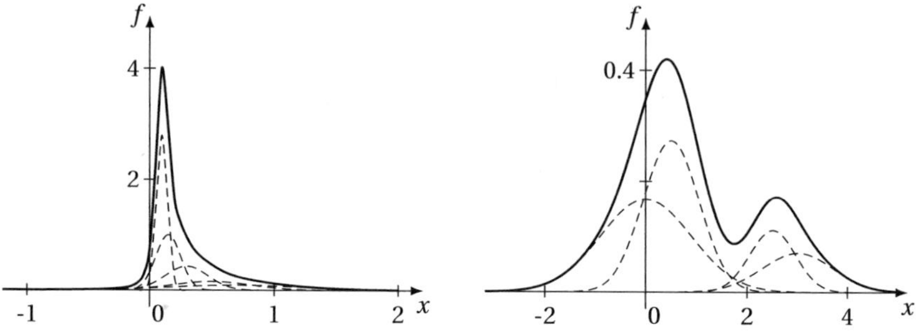

(a) Heavily skewed density consisting of five components.

(b) Multimodal density consisting of four components.

Figure 3.1: Two exemplary Gaussian mixture densities. The individual Gaussian components are plotted as dashed lines.

3.1 Gaussian Mixtures

A Gaussian mixture density is defined as a weighted sum of Gaussian densities according to

$$f(\underline{x}) = \sum_{i=1}^{L} \omega_i \cdot \mathcal{N}(\underline{x}; \underline{\mu}_i, \mathbf{C}_i) \tag{3.1}$$

with non-negative weighting coefficients ω_i that sum up to one. In Figure 3.1, some examples of Gaussian mixture densities are shown. Sometimes this density type is also called Gaussian sum density or mixture of Gaussians. Obviously, the Gaussian density is a special case of (3.1) for $L = 1$. By defining the two vectors $\underline{\omega} \triangleq [\omega_1 \; \dots \; \omega_L]^{\mathrm{T}}$ and $\underline{f}(\underline{x}) \triangleq [\mathcal{N}(\underline{x}; \underline{\mu}_1, \mathbf{C}_1) \; \dots \; \mathcal{N}(\underline{x}; \underline{\mu}_L, \mathbf{C}_L)]^{\mathrm{T}}$, a more compact version of (3.1) can be found via

$$f(\underline{x}) = \underline{\omega}^{\mathrm{T}} \cdot \underline{f}(\underline{x}) \,. \tag{3.2}$$

There are mainly three reasons for the widespread use of Gaussian mixtures in Bayesian filtering: First, as they consist of multiple Gaussians, they possess a straightforward parametrization by means of the weights ω_i, mean vectors $\underline{\mu}_i$ and covariance matrices $\mathbf{C}_i$ of the individual Gaussian components. Second, the practically relevant moments mean and covariance of the Gaussian mixture can be calculated in closed form by means of

$$\underline{\mu}_x = \sum_{i=1}^{L} \omega_i \cdot \underline{\mu}_i \,,$$

$$\mathbf{C}_x = \sum_{i=1}^{L} \omega_i \cdot \left(\mathbf{C}_i + \underline{\mu}_i \cdot \underline{\mu}_i^{\mathrm{T}} \right) - \underline{\mu}_x \cdot \underline{\mu}_x^{\mathrm{T}} \,.$$

And finally, any continuous density function $\tilde{f}(\underline{x})$ can be approximated by means of a Gaussian mixture as closely as required with respect to the *Lissack-Fu* distance

$$\int \left| \tilde{f}(\underline{x}) - f(\underline{x}) \right| \mathrm{d}\underline{x}$$

by increasing the number of components L and when $\mathbf{C}_i$ approaches the zero matrix [7, 181]. Thus, a Gaussian mixture density can be considered as *universal approximator* for density functions [122].

3.2 Nonlinear Filtering

For a Gaussian mixture filter it is assumed that both the predicted and the posterior density of the state are represented as Gaussian mixtures according to

$$f_k^{\bullet}(\underline{x}_k) = \sum_{i=1}^{L_k^{\bullet}} \omega_{k,i}^{\bullet} \cdot \mathcal{N}\left(\underline{x}_k ; \underline{\hat{x}}_{k,i}^{\bullet}, \mathbf{C}_{k,i}^{\bullet} \right) \,, \text{ with } \bullet \in \{e,p\} \,, \tag{3.3}$$

for every time step k, where $L_k^{\bullet}$ is the number of mixture components. To calculate the parameters of these densities, several approaches exist that can be grouped in three major classes as depicted in Figure 3.2. In *model approximating* approaches, the transition density (1.6) and the

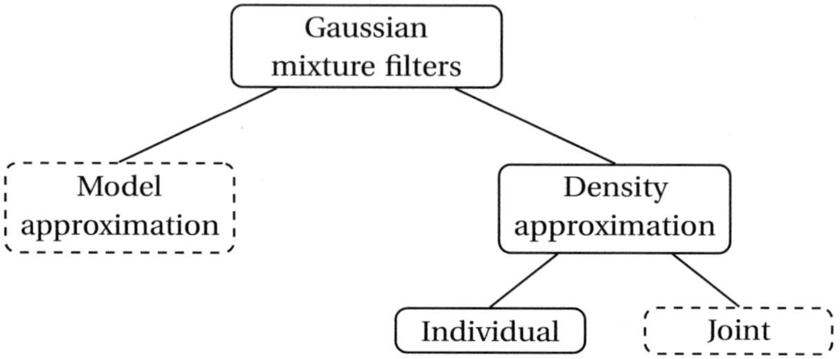

Figure 3.2: Taxonomy for Gaussian mixture filters. The dashed classes are not considered in this thesis.

likelihood (1.8) are approximated by means of a Gaussian mixture, which is typically done off-line, before the actual filtering. For instance [84, 133] propose techniques for approximating the transition density, while likelihood approximation is content of [5, 87, 192].

Density approximation approaches instead focus on directly approximating the true predicted or posterior density by means of Gaussian mixtures. Here, one can distinguish between *joint approximation*, where all Gaussian components and their respective parameters are calculated jointly, typically by solving an optimization problem as proposed in [76, 77]. In case of *individual approximations*, each Gaussian component is processed separately through the prediction and measurement update.

3.2.1 Individual Approximation

In this thesis, the focus is on individual approximation techniques as they allow a direct utilization of the Gaussian filtering algorithms discussed in the previous chapter. In doing so, the calculation of the desired parameters of the predicted and posterior Gaussian mixtures can be performed efficiently without any demanding off-line approximations. In the following, the prediction and measurement update of a generic Gaussian mixture filter based on individual approximation are derived.

Prediction

For calculating the predicted density, the Gaussian mixture representing the posterior distribution $f_k^e(\underline{x}_k)$ is plugged into (1.5), which yields

$$\begin{aligned} f_{k+1}^p(\underline{x}_{k+1}) &= \int f(\underline{x}_{k+1}|\underline{x}_k,\underline{u}_k)\cdot\left(\sum_{i=1}^{L_k^e}\omega_{k,i}^e\cdot\mathcal{N}\left(\underline{x}_k;\hat{\underline{x}}_{k,i}^e,\mathbf{C}_{k,i}^e\right)\right)\mathrm{d}\underline{x}_k \\ &\approx \sum_{i=1}^{L_k^e}\underbrace{\omega_{k,i}^e}_{\equiv\,\omega_{k+1,i}^p}\cdot\underbrace{\int f(\underline{x}_{k+1}|\underline{x}_k,\underline{u}_k)\cdot\mathcal{N}\left(\underline{x}_k;\hat{\underline{x}}_{k,i}^e,\mathbf{C}_{k,i}^e\right)\mathrm{d}\underline{x}_k}_{\approx\,\mathcal{N}\left(\underline{x}_{k+1};\underline{\mu}_{k+1,i}^p,\mathbf{C}_{k+1,i}^p\right)}\ . \end{aligned} \tag{3.4}$$

with $L_k^p \equiv L_k^e$. The integral cannot be solved in closed form except for the linear case. To simplify the integration, for each individual component of the posterior mixture the solution of the integral is approximated by means of the Gaussian $\mathcal{N}(\underline{x}_{k+1};\underline{\mu}_{k+1,i}^p,\mathbf{C}_{k+1,i}^p)$. Thus, it remains to calculate the mean vector $\underline{\mu}_{k+1,i}^p$ and covariance matrix $\mathbf{C}_{k+1,i}^p$. For this purpose any of the Gaussian filters of Chapter 2 can be employed. The weights remain unchanged. The resulting predicted density $f_{k+1}^p(\underline{x}_{k+1})$ is then again a Gaussian mixture.

Measurement Update

In case of the measurement update, the predicted mixture (3.4) is substituted in (1.7) according to

$$f_k^e(\underline{x}_k) = c_k\cdot f(\hat{\underline{z}}_k|\underline{x}_k)\cdot\left(\sum_{i=1}^{L_k^p}\omega_{k,i}^p\cdot\mathcal{N}\left(\underline{x}_k;\underline{\mu}_{k,i}^p,\mathbf{C}_{k,i}^p\right)\right) \tag{3.5}$$

Due to the nonlinearity of the measurement equation, the measurement update cannot be solved analytically, i.e., the normalization constant as well as the product of the likelihood with the predicted mixture possess no closed-form expression in general.

The normalization constant $c_k = {}^1/_{f(\hat{\underline{z}}_k)}$ corresponds to the reciprocal probability of the measurement. This probability can be approximated as in the prediction step according to

$$
\begin{aligned}
f(\hat{\underline{z}}_k) = \frac{1}{c_k} &= \int f(\hat{\underline{z}}_k | \underline{x}_k) \cdot \left(\sum_{i=1}^{L_k^p} \omega_{k,i}^p \cdot \mathcal{N}\left(\underline{x}_k; \underline{\mu}_{k,i}^p, \mathbf{C}_{k,i}^p\right) \right) \mathrm{d}\underline{x}_k \\
&\approx \sum_{i=1}^{L_k^p} \omega_{k,i}^p \cdot \mathcal{N}\left(\hat{\underline{z}}_k; \underline{\mu}_{k,i}^z, \mathbf{C}_{k,i}^z\right) ,
\end{aligned} \tag{3.6}
$$

where the parameters $\underline{\mu}_{k,i}^z$, $\mathbf{C}_{k,i}^z$ are determined individually by means of a Gaussian filter.

To approximate the product between likelihood and predicted mixture in (3.5), the equation is extended with the components $\mathcal{N}(\hat{\underline{z}}_k; \underline{\mu}_{k,i}^z, \mathbf{C}_{k,i}^z)$ from (3.6), which results in

$$
f_k^e(\underline{x}_k) \approx \sum_{i=1}^{L_k^p} \underbrace{c_k \cdot \omega_{k,i}^p \cdot \mathcal{N}\left(\hat{\underline{z}}_k; \underline{\mu}_{k,i}^z, \mathbf{C}_{k,i}^z\right)}_{\triangleq\, \omega_{k,i}^e} \cdot \underbrace{\frac{f(\hat{\underline{z}}_k | \underline{x}_k) \cdot \mathcal{N}\left(\underline{x}_k; \underline{\mu}_{k,i}^p, \mathbf{C}_{k,i}^p\right)}{\mathcal{N}\left(\hat{\underline{z}}; \underline{\mu}_{k,i}^z, \mathbf{C}_{k,i}^z\right)}}_{\approx\, \mathcal{N}\left(\underline{x}_k; \underline{\mu}_{k,i}^e, \mathbf{C}_{k,i}^e\right)} \tag{3.7}
$$

with $L_k^p \equiv L_k^e$. The fraction in (3.7) corresponds to a Bayesian measurement update for each individual predicted Gaussian and is approximated with a Gaussian density $\mathcal{N}(\underline{x}_k; \underline{\mu}_{k,i}^e, \mathbf{C}_{k,i}^e)$. The mean vector and covariance matrix of these individual posterior Gaussian components are determined by means of a Gaussian filter by employing the conditioning in (2.9). Therefore, the parameters $\underline{\mu}_{k,i}^z$, $\mathbf{C}_{k,i}^z$ are required, which are already available from (3.6). Only the cross-covariance matrices $\mathbf{C}_{k,i}^{xz}$ need to be calculated in addition.

3.2.2 Generic Gaussian Mixture Filter

Thanks to the individual approximation, both the prediction and measurement update of the Gaussian mixture filter boil down to a *bank of Gaussian filters*, where each individual Gaussian filter tracks the evolution of its assigned Gaussian component. The Gaussian mixture in each estimation step is the linear combination of the individual results [74].

Besides the pure estimation steps, a generic Gaussian mixture filter requires additional operations for maintaining an accurate density approximation and an adequate run-time. In Algorithm 3, these additional operations are listed for both prediction and measurement update. The *refinement* replaces components of the given Gaussian mixture with one or more new components. This might be necessary to overcome strong nonlinearities in the system model or measurement model. As all Gaussian filters either explicitly or implicitly perform a linearization, the linearization error can be reduced by refining components [3, 64, 146].

The *reapproximation* comprises two operations: weight optimization and reduction. The individual weights of the posterior Gaussian components in (3.7) are merely an approximation of the true weights. This observation follows from the approximate calculation of the normalization in (3.6). The normalization is one factor forming the posterior weights. To improve the weights, on additional *weight optimization* can be performed after the measurement update.

The *reduction* step removes Gaussian components from the predicted and posterior mixture. The need for this operation has many reasons: Components may become negligible due to very low weights. Furthermore, due to the refinement, the number of mixture components grows over time. This growth will become a severe problem, when in addition the noise

Algorithm 3 Generic Gaussian mixture filter

Initialize state density $f_0(\underline{x})$ with a Gaussian mixture
for each time step k **do**
  ▷ *Prediction*
  `Refinement:` Introduce additional components
  `Estimation:` Compute predicted Gaussian mixture $f_k^p(\underline{x}_k)$
  `Reapproximation`: Weight optimization & component reduction
  ▷ *Measurement Update*
  `Refinement:` Introduce additional components
  `Estimation:` Compute posterior Gaussian mixture $f_k^e(\underline{x}_k)$
  `Reapproximation`: Weight optimization & component reduction
end for

components $\underline{\boldsymbol{w}}_k$ and $\underline{\boldsymbol{v}}_k$ are itself represented as Gaussian mixtures[1]. As the noise terms form the transition density and likelihood, respectively, the multiplication of $f(\underline{x}_{k+1}|\underline{x}_k, \underline{u}_k)$ with $f_k^e(\underline{x}_k)$ in the prediction step (3.4) and the multiplication of $f(\underline{\hat{z}}_k|\underline{x})$ with $f_k^p(\underline{x}_k)$ in the measurement update (3.7) will lead to an exponential growth of the components. Here, the reduction is a must to maintain a feasible algorithm.

3.3 Component Adaptation

In this section, a detailed overview of the refinement and reapproximation operations appearing in Algorithm 3. At first, the two sub-operations weight optimization and reduction of the reapproximation step are discussed. The weight optimization part is kept short as it is not considered

1 Even if the noise is Gaussian, an approximation of the noise by means of a Gaussian mixture might be reasonable if the noise covariance is large. In case of a large covariance, the individual processing considered above becomes prone to large linearization errors. An approximation of the noise by a Gaussian mixture leads to components with lower covariances [7].

further in this thesis and thus, it is merely mentioned for the sake of completeness.

3.3.1 Weight Optimization

The posterior weight calculation in (3.7) facilitates the individual processing of the Gaussian components, but it is merely approximate as discussed above. To minimize the deviation between the mixture and the true posterior density, [91] proposed to adjust the weights by minimizing the norm

$$\int \left\| f_k^e(\underline{x}_k) - \sum_{i=1}^{L_k^e} \omega_{k,i}^e \cdot \mathcal{N}\left(\underline{x}_k; \underline{\mu}_{k,i}^e, \mathbf{C}_{k,i}^e\right) \right\| \mathrm{d}\underline{x}_k \,. \tag{3.8}$$

As the true posterior $f_k^e(\underline{x}_k)$ is not known and cannot be calculated analytically, this norm is evaluated at so-called *collocation points*. A natural choice of these collocation points are the mean vectors $\underline{\mu}_{k,i}^e$ of the posterior mixture. By this choice of collocation points and by using the L_2 norm in (3.8), the weight adjustment becomes a least squares optimization problem.

At a first glance, the weights obtained after the prediction seem to be correct. No approximation with respect to the weights is involved. However, no update of the weights is performed in the prediction; they coincide with the posterior weights. This procedure is optimal only for linear models, but becomes suboptimal when performing individual linearizations by means of a bank of Gaussian filters, especially when the covariances of the individual components is large and thus, the components may have a large overlap. For this situation, [194, 195] proposed a weight optimization procedure for the prediction step similar to the above posterior weight optimization. Instead of collocation points to evaluate (3.8), [194, 195] replace the true predicted density with the approximate posterior mixture density of the previous measurement update.

3.3.2 Reduction

To bound the growth of the number of Gaussian mixture components, *mixture reduction* aims at replacing a given Gaussian mixture $f(\underline{x})$ with L components as in (3.1) by a mixture $\tilde{f}(\underline{x})$ with M components, where M is significantly smaller than L, i.e., $M \ll L$. At the same time, the deviation between the true and the reduced mixture should kept at a minimum.

Deviation Measures

A very natural deviation measure between two densities from an information theoretic perspective is the *Kullback-Leibler divergence* (KLD)

$$G\left(f(\underline{x}) \| \tilde{f}(\underline{x})\right) \triangleq \int f(\underline{x}) \cdot \log \frac{f(\underline{x})}{\tilde{f}(\underline{x})} \, \mathrm{d}\underline{x} \,. \tag{3.9}$$

It can be interpreted as a quantification of the likelihood that data drawn from the true density is spuriously considered as be drawn from the reduced mixture and thus, it measures the difficulty of discriminating two densities. As discussed in [156, 211], thanks to this maximum likelihood interpretation and its scale-independence, the KLD should be the preferred choice for Gaussian mixture reduction. Unfortunately, the KLD is not a distance measure or norm in a strict sense as it is not symmetric. Furthermore, due to the logarithm in (3.9), the KLD cannot be evaluated in closed form for Gaussian mixtures.

To overcome the restrictions of the KLD, the *integrated squared difference* (ISD) is often employed as an alternative. It is defined as

$$D\left(f(\underline{x}), \tilde{f}(\underline{x})\right) \triangleq \int \left(f(\underline{x}) - \tilde{f}(\underline{x})\right)^2 \mathrm{d}\underline{x} \,. \tag{3.10}$$

Thus, the ISD is actually an L_2 norm. In contrast to the KLD, it has a closed-form expression for Gaussian mixtures, which is given by

$$
\begin{aligned}
D\big(f(\underline{x}), \tilde{f}(\underline{x})\big) = & \sum_{i=1}^{L}\sum_{j=1}^{L} \omega_i \cdot \omega_j \cdot \mathcal{N}\Big(\underline{\mu}_i; \underline{\mu}_j, \mathbf{C}_i + \mathbf{C}_j\Big) + \\
& \sum_{i=1}^{L}\sum_{j=1}^{M} \omega_i \cdot \tilde{\omega}_j \cdot \mathcal{N}\Big(\underline{\mu}_i; \underline{\tilde{\mu}}_j, \mathbf{C}_i + \tilde{\mathbf{C}}_j\Big) + \\
& \sum_{i=1}^{M}\sum_{j=1}^{M} \tilde{\omega}_i \cdot \tilde{\omega}_j \cdot \mathcal{N}\Big(\underline{\tilde{\mu}}_i; \underline{\tilde{\mu}}_j, \tilde{\mathbf{C}}_i + \tilde{\mathbf{C}}_j\Big) ,
\end{aligned}
\tag{3.11}
$$

where the first term is called *self-similarity* of the true mixture, the second is the *cross-similarity* between the true and the reduced mixture, and the last term is the *self-similarity* of the reduced mixture [211].

State-of-the-Art

Depending on the optimization that is performed by a Gaussian mixture reduction algorithm, one can distinguish three basic classes: local, global, and pseudo-global algorithms. In the following, the key features and typical approaches of every class are discussed.

Local reduction algorithms typically perform an iterative *merging* of two or more components to a single Gaussian until a pre-defined threshold on the maximum allowed number of components is reached.

Example 11: Moment-preserving Merge

Let $\omega_i \cdot \mathcal{N}(\underline{x}; \underline{\mu}_i, \mathbf{C}_i)$ and $\omega_j \cdot \mathcal{N}(\underline{x}; \underline{\mu}_j, \mathbf{C}_j)$ be the two Gaussians that are considered for merging. By merging these two components, it should in addition be ensured that the zeroth-order (probability mass), first-order (mean) and second-order (covariance) moments of the entire mixture remain unchanged. Given these constraints, a *moment-preserving merge* of the two Gaussians yields the parameters

$$
\begin{aligned}
\tilde{\omega} &= \omega_i + \omega_j \,, \\
\underline{\tilde{\mu}} &= \tfrac{1}{\tilde{\omega}} \cdot \left(\omega_i \cdot \underline{\mu}_i + \omega_j \cdot \underline{\mu}_j \right) , \\
\tilde{\mathbf{C}} &= \tfrac{1}{\tilde{\omega}} \cdot \left(\omega_i \cdot \mathbf{C}_i + \omega_j \cdot \mathbf{C}_j + \tfrac{\omega_i \cdot \omega_j}{\tilde{\omega}} \cdot \left(\underline{\mu}_i - \underline{\mu}_j \right) \left(\underline{\mu}_i - \underline{\mu}_j \right)^{\mathrm{T}} \right) .
\end{aligned}
\tag{3.12}
$$

of the resulting single Gaussian $\tilde{\omega} \cdot \mathcal{N}(\underline{x}; \underline{\tilde{\mu}}, \tilde{\mathbf{C}})$.

The above moment-preserving merge can be easily extended for merging more than two Gaussians. The major difference of the most local algorithms is the criterion based on which components are selected for a merge. In [136, 158, 208] for instance, the Mahalanobis distance between Gaussians is utilized, while [189] employes a pair-wise version of the Hellinger metric and [40] a pair-wise version of the ISD. All these criteria only measure some similarity between components without consideration of the induced global deviation between the true and reduced mixture due to the merge, i.e., neither the KLD nor the ISD are monitored. Certainly, local algorithms are well-suited for real-time applications due to their low computational overhead.

In contrast, *global reduction* algorithms directly aim at minimizing the KLD or the ISD. In doing so, this class of algorithms tries to find the globally optimal reduced mixture. The reduction method proposed in [90] optimizes the parameters of the reduced mixture with respect to the ISD. This approach is constructive as it begins with a single Gaussian and adds components at locations where the deviation between the true and reduced mixture is too large. The algorithms in [31, 202] utilize the KLD, where the first is an expectation-maximization like approach and the second relies on variational Bayes. Merging components—when not performed iteratively as done by the local algorithms—can also yield globally optimal results as demonstrated in [47]. Here, all possible merges are first calculated and then the optimal solution is selected, which directly gives the reduced mixture. This procedure, however, is only feasible for L being a low number. In general, global algorithms suffer from a high computational load.

A compromise between both worlds is the third class of *pseudo-global* reduction algorithms. Here, iterative merging is performed as in local algorithms, but at the same time the KLD or ISD is monitored. [211] for instance considers that pair of Gaussians for a merge that introduces the smallest error into the reduced mixture in an ISD sense. [156] instead employes an upper-bound of the KLD and [168] uses only the cross-likeliness of the ISD. Because every possibly merge is inspected before actually performing the merge, pair-wise merging is globally optimal in a single step. Though, this does not guarantee global optimality after multiple reduction steps. Global optimality is only achieved by the aforementioned procedure proposed in [47]. To compensate this drawback, some algorithms like those in [70, 163] perform a dedicated optimization after merging, where the parameters of the reduced mixture are optimized with respect to the ISD. Alternatively to pair-wise merging, clustering-based or k-means based reductions perform iterated assignments of components of the true mixture to clusters. When no improvement in terms of the KLD or ISD is gained anymore, the components assigned to a cluster a merged which yields the reduced mixture.

Open Issues

Most of the aforementioned algorithms require a pre-defined threshold on the resulting number of mixture components, where an appropriate choice is difficult for the user. If the threshold is chosen too high, the entire Gaussian mixture filter becomes computationally demanding. A too low threshold may lead to poor estimates or even to a diverging filter. Ideally, this so-called *model selection* problem should be solved by the reduction algorithm itself, which so far is only achieved by [31, 90].

Both algorithms, however, are computationally and algorithmically very complex. Reductions of these algorithms and also of other global and pseudo-global approaches, are often more demanding than the actual predictions and measurement updates.

3.3.3 Refinement

While reduction aims at removing components from the mixture, the refinement step introduces new ones. The refinement becomes necessary when severe linearization errors of the individual Gaussian filters threaten the estimation performance.

Nonlinearity Measures

In order to avoid a blind adding of new components, the potential linearization error or the "strength" of the nonlinearity needs to be quantified. For some Gaussian filters, dedicated *nonlinearity measures* have been proposed in the past. For a brief overview, the nonlinear transformation in (2.6) is revisited, where now $\underline{\boldsymbol{x}}$ and $\underline{\boldsymbol{y}}$ are Gaussian mixtures as in (3.1).

If for instance an EKF is employed as Gaussian filter, [3, 93] propose the measure

$$N\left(\underline{\mu}_i, \mathbf{C}_i\right) \triangleq \frac{\sqrt{\operatorname{Tr}\left(\mathbf{G}_{xx}\left(\underline{\mu}_i\right)\mathbf{C}_i\mathbf{G}_{xx}\left(\underline{\mu}_i\right)\mathbf{C}_i\right)}}{\sigma_w} ,$$

where $\mathbf{G}_{xx}\left(\underline{x}\right) \triangleq \partial^2 g(\underline{x})/\partial\underline{x}\partial\underline{x}^{\mathrm{T}}$ is the Hessian matrix of $\underline{g}(.)$. Unfortunately, this measure is restricted to EKFs and to scalar $\boldsymbol{y}$ only. A measure based on the KLD (3.9) that is not limited by the dimension of $\boldsymbol{y}$ is proposed in [146]. Here, every mixture component of $\underline{\boldsymbol{y}}$ is compared against the true density. As the true density is not given in closed-form, numerical integration is necessary, which is demanding for high dimensions.

For the UKF, [63] proposed the measure

$$\sum_{i=1}^{n_y} \eta_i \,, \quad \text{with } \eta_i \triangleq \tfrac{1}{2} \left\| \mathcal{Y}_i + \mathcal{Y}_{n_x+i} - \mathcal{Y}_{2n_x+1} \right\|_2^2$$

that quantifies the goodness of fit of the propagated sample points $\mathcal{Y}_i = \underline{g}(\mathcal{X}_i)$, $i = 1 \ldots n_x$, to a linear regression model. The measure is close to zero if $\underline{g}(.)$ is approximately linear.

Adding New Components

By means of the nonlinearity measures it is possible to locate the Gaussian at the strongest nonlinearity or linearization error. A typical action to attenuate the linearization error is *splitting*, i.e., the identified Gaussian component $\omega \cdot \mathcal{N}(\underline{x}; \underline{\mu}, \mathbf{C})$ is replaced by a Gaussian mixture according to

$$\omega \cdot \mathcal{N}\left(\underline{x}; \underline{\mu}, \mathbf{C}\right) \approx \sum_{i=1}^{L} \omega_i \cdot \mathcal{N}\left(\underline{x}; \underline{\mu}_i, \mathbf{C}_i\right) , \tag{3.13}$$

where the components on the right-hand side possess smaller covariances than the original component on the left-hand side, which is necessary for reducing the linearization error.

It is worth mentioning that (3.13) has two interpretations. Reading the equation from left to right corresponds to splitting, but reading the equation from right to left is nothing else then mixture reduction. Hence, splitting can be considered the dual operation to mixture reduction.

It can be easily verified that for $L > 1$, the number of free parameters on the right-hand side of (3.13), i.e., weights, mean vectors, and covariance matrices, is larger than the number of given parameters. Hence, splitting a Gaussian is an *ill-posed problem*.

The solution to this problem proposed in [6] is based on the UKF for determining sample points of the Gaussian that has to be split. The sample points with corresponding weights are used as the mean vectors

$\underline{\mu}_i$ and mixture weights ω_i in (3.13), respectively. The covariances $\mathbf{C}_i$ are chosen to be identical. In [7, 63], the new Gaussians are placed as a regular grid with identical covariance matrices. The weights are chosen to be $\omega_i = \mathcal{N}\left(\underline{\mu}_i; \underline{\mu}, \mathbf{C}\right)$, i.e., the normalized probability value of the original Gaussian at the means $\underline{\mu}_i$ of the new Gaussian. A so-called splitting library is used in [76, 86], i.e., an off-line calculated approximation of a standard Gaussian by a mixture of Gaussians. This approximation is adjusted on-line by means of straightforward scaling operations.

Open Issues

The discussed nonlinearity measures are either restricted to very specific Gaussian filters like the UKF and EKF or they are difficult to evaluate for arbitrary models. Except of the splitting library all splitting approaches suffer from scalability problems as the number of components scales with the dimension of the state space. Furthermore, splitting is performed very "generous", i.e., new components are introduced at every dimension of the state space, although the linearization error might affect only some state variables.

3.4 Contributions

The contributions made by the Papers E–G are discussed in the following sections. At first, the nonlinear-nonlinear state decomposition approach that was proposed in Section 2.5.2 is extended to Gaussian mixture filters. In Section 3.4.2, a new measure of the degree of nonlinearity and a new splitting approach are proposed. Both together form the so-called *adaptive Gaussian mixture filter*. Finally, a global Gaussian mixture reduction algorithm that exploits the ISD and the curvature of the true density is introduced.

3.4.1 Semi-Analytic Gaussian Mixture Filter

As the generic Gaussian mixture filter in Algorithm 3 uses a bank of Gaussian filters, the SAGF proposed in Section 2.5.2 can be directly applied in order to obtain a *semi-analytic Gaussian mixture filter* (SAGMF). As in Section 2.5.2, the conditionally integrable nonlinear transformation

$$\underline{y} = \underline{g}(\underline{x}_a, \underline{x}_s) + \underline{w}$$

is considered, where the state $\underline{x}$ comprises the conditionally integrable state $\underline{x}_a$ and the sampled state $\underline{x}_s$. The density of the state is

$$f(\underline{x}) = \sum_{i=1}^{L} \omega_i \cdot \mathcal{N}\left(\underline{x}; \underline{\mu}_i^x, \mathbf{C}_i^x\right) \quad \text{with} \quad \underline{\mu}_i^x = \begin{bmatrix} \underline{\mu}_i^a \\ \underline{\mu}_i^s \end{bmatrix}, \mathbf{C}_i^x = \begin{bmatrix} \mathbf{C}_i^a & \mathbf{C}_i^{as} \\ \mathbf{C}_i^{sa} & \mathbf{C}_i^s \end{bmatrix}. \tag{3.14}$$

To obtain the Gaussian mixture density $\sum_{i=1}^{L} \omega_i \cdot \mathcal{N}(\underline{y}; \underline{\mu}_i^y, \mathbf{C}_i^y)$ of $\underline{y}$, it is necessary to solve the moment integrals (2.8) for every component of (3.14). At first, for every component $i = 1 \ldots L$, a sample approximation of the marginal Gaussian $\mathcal{N}(\underline{x}_s; \underline{\mu}_i^s, \mathbf{C}_i^s)$ is calculated by means of an LRKF, which yields

$$\mathcal{N}\left(\underline{x}_s; \underline{\mu}_i^s, \mathbf{C}_i^s\right) \approx \sum_{j=1}^{N} \omega_{ij} \cdot \delta\left(\underline{x}^s - \mathcal{X}_{ij}\right),$$

where $\mathcal{L} = \{\omega_{ij}, \mathcal{X}_{ij}\}$ for $j = 1 \ldots N$ are the sample points of the ith Gaussian component.

Based on this sample representation, it is now possible to determine the mean and covariance of the ith component of $\underline{y}$ according to

$$\underline{\mu}_i^y \approx \sum_{j=1}^{N} \omega_{ij} \cdot \underline{\mu}_{ij}^y, \tag{3.15}$$

$$\mathbf{C}_i^y \approx \sum_{j=1}^{N} \omega_{ij} \cdot \left(\mathbf{C}_{ij}^y - \underline{\mu}_{ij}^y \left(\underline{\mu}_i^y\right)^{\mathrm{T}} - \underline{\mu}_i^y \left(\underline{\mu}_{ij}^y\right)^{\mathrm{T}} + \underline{\mu}_i^y \left(\underline{\mu}_i^y\right)^{\mathrm{T}}\right), \tag{3.16}$$

with

$$\underline{\mu}_{ij}^{y} = \int \underline{g}\left(\underline{x}_a, \mathcal{X}_{ij}\right) \cdot f_i\left(\underline{x}_a | \mathcal{X}_{ij}\right) \mathrm{d}\underline{x}_a \, ,$$
$$\mathbf{C}_{ij}^{y} = \int \underline{g}\left(\underline{x}_a, \mathcal{X}_{ij}\right) \cdot \underline{g}\left(\underline{x}_a, \mathcal{X}_{ij}\right)^{\mathrm{T}} \cdot f_i\left(\underline{x}_a | \mathcal{X}_{ij}\right) \mathrm{d}\underline{x}_a \, , \tag{3.17}$$

where $f_i\left(\underline{x}_a | \mathcal{X}_{ij}\right) \triangleq \mathcal{N}\left(\underline{x}_a; \underline{\mu}_i^{a|s}, \mathbf{C}_i^{a|s}\right)$ is the conditional density of the ith component with mean and covariance as in (2.48). As $\underline{g}(.)$ is conditionally integrable, the integrals in (3.17) can be solved analytically.

To obtain the cross-covariance $\mathbf{C}_i^{xy} = \left[\mathbf{C}_i^{ay} \quad \mathbf{C}_i^{sy}\right]^{\mathrm{T}}$ it is more convenient to calculate its sub-matrices separately according to

$$\mathbf{C}_i^{ay} = \sum_{j=1}^{N} \omega_{ij} \cdot \left(\mathbf{C}_{ij}^{ay} - \underline{\mu}_{ij}^{a|s}\left(\underline{\mu}_i^{y}\right)^{\mathrm{T}} + \underline{\mu}_i^{a}\left(\underline{\mu}_i^{y} - \underline{\mu}_{ij}^{y}\right)^{\mathrm{T}}\right) \, ,$$
$$\mathbf{C}_i^{sy} = \sum_{j=1}^{N} \omega_{ij} \cdot \left(\mathcal{X}_{ij} - \underline{\mu}_i^{s}\right) \cdot \left(\underline{\mu}_{ij}^{y} - \underline{\mu}_i^{y}\right)^{\mathrm{T}} \, , \tag{3.18}$$

with

$$\mathbf{C}_{ij}^{ay} = \int \underline{x}_a \cdot \underline{g}\left(\underline{x}_a, \mathcal{X}_{ij}\right)^{\mathrm{T}} \cdot f_i\left(\underline{x}_a | \mathcal{X}_{ij}\right) \mathrm{d}\underline{x}_a \, ,$$

where $\underline{\mu}_{ij}^{y}$ is given by (3.17) and $\underline{\mu}_{ij}^{a|s}$ is calculated according to (2.48) with $\underline{x}_s$ being replaced by $\mathcal{X}_{ij}$.

With the mean vectors (3.15), covariance matrices (3.16), and the sub-matrices (3.18) of the cross-covariance matrix all ingredients are given that are necessary for calculating the individual components of the predicted and posterior Gaussian mixture densities.

Example 12: Tricycle Kinematics

A robot with tricycle kinematics has to be localized. The nonlinear kinematics model is defined as

$$\begin{aligned}\boldsymbol{p}_{k+1}^{x} &= \boldsymbol{p}_{k}^{x} + \left(u_{k}^{v} + \underline{\boldsymbol{w}}_{k}^{v}\right) \cdot \cos\left(\boldsymbol{\phi}_{k} + u_{k}^{\alpha}\right) , \\ \boldsymbol{p}_{k+1}^{y} &= \boldsymbol{p}_{k}^{y} + \left(u_{k}^{v} + \underline{\boldsymbol{w}}_{k}^{v}\right) \cdot \sin\left(\boldsymbol{\phi}_{k} + u_{k}^{\alpha}\right) , \\ \boldsymbol{\phi}_{k+1} &= \boldsymbol{\phi}_{k} + \left(u_{k}^{\alpha} + \underline{\boldsymbol{w}}_{k}^{\alpha}\right) ,\end{aligned}$$

with state $\underline{\boldsymbol{x}}_k = \left[\boldsymbol{p}_k^x \ \boldsymbol{p}_k^y \ \boldsymbol{\phi}_k\right]^{\mathrm{T}}$, where $\boldsymbol{p}_k^x$ and $\boldsymbol{p}_k^y$ describe the Cartesian position of the robot and $\boldsymbol{\phi}_k$ its orientation. The initial position of the robot at time step $k = 0$ is $\underline{x}_0 = [5\ 3\ 0.2]^{\mathrm{T}}$. The known control inputs are the velocity $u_k^v = 0.1$ and the turning angle $u_k^\alpha = 0.1$. $\underline{\boldsymbol{w}}_k^v$ and $\underline{\boldsymbol{w}}_k^\alpha$ are noise processes affecting the corresponding control inputs. They are assumed to be zero-mean Gaussian with variances $\sigma_{wv}^2 = 0.1$ and $\sigma_{w\alpha}^2 = 0.01$, respectively.

The state is decomposed into $\underline{\boldsymbol{x}}^a = \left[\boldsymbol{p}^x \ \boldsymbol{p}^y \ \underline{\boldsymbol{w}}^v \ \underline{\boldsymbol{w}}^\alpha\right]^{\mathrm{T}}$ and $\boldsymbol{x}^b = \boldsymbol{\phi}$. Hence, by conditioning on $\boldsymbol{x}_s$, the system model becomes linear and can be solved via the prediction of the Kalman filter.

Range measurements to landmarks are performed for localizing the robot. The measured range $\boldsymbol{r}_k$ is defined by the nonlinear measurement model

$$\boldsymbol{r}_k = \sqrt{\left(\boldsymbol{p}_k^x - L^x + \boldsymbol{v}_k^x\right)^2 + \left(\boldsymbol{p}_k^y - L^y + \boldsymbol{v}_k^y\right)^2} ,$$

where $\underline{L} = \left[L^x \ L^y\right]^{\mathrm{T}}$ is the position of the landmark. Four landmarks at the positions

$$\left[\underline{L}_1 \quad \underline{L}_3 \quad \underline{L}_3 \quad \underline{L}_4\right] = \begin{bmatrix} 0 & 2 & 5 & 10 \\ 0 & 2 & 5 & 10 \end{bmatrix} .$$

are given. The measurement noise $\underline{\boldsymbol{v}}_k = \left[\boldsymbol{v}_k^x \ \boldsymbol{v}_k^y\right]^{\mathrm{T}}$ is zero-mean Gaussian with covariance $\mathbf{C}^v = \sigma_v^2 \cdot \mathbf{I}_2$. For the variance σ_v^2, the three noise

Table 3.1: Average rmse and standard deviation for the different filters at different noise levels and numbers of components.

Noise 0.5	64	8	1
FAGMF	**2.02 ± 1.34**	**3.17 ± 1.98**	**3.76 ± 2.01**
SAGMF	2.03 ± 1.34	**3.16 ± 1.99**	3.76 ± 2.09
UGMF	2.41 ± 1.79	4.08 ± 3.02	4.31 ± 3.58
EGMF	2.64 ± 1.86	4.08 ± 4.99	6.40 ± 9.64

Noise 1.0	64	8	1
FAGMF	**2.05 ± 1.36**	3.16 ± 1.97	**3.70 ± 2.07**
SAGMF	**2.06 ± 1.35**	**3.15 ± 1.97**	**3.70 ± 2.07**
UGMF	2.44 ± 1.80	4.04 ± 3.05	4.25 ± 3.61
EGMF	2.95 ± 1.73	4.40 ± 4.43	6.72 ± 8.71

Noise 2.0	64	8	1
FAGMF	**2.16 ± 1.35**	**3.22 ± 2.02**	**3.78 ± 2.16**
SAGMF	2.16 ± 1.36	**3.22 ± 2.02**	**3.78 ± 2.16**
UGMF	2.61 ± 1.77	4.11 ± 3.08	4.35 ± 3.65
EGMF	3.42 ± 1.55	4.95 ± 4.14	7.21 ± 7.60

levels 0.5, 1, and 2 are considered. The measurement step is performed analytically by means of the closed-form solution derived in Section 5.1.

The proposed semi-analytic Gaussian mixture filter (SAGMF) is compared against Gaussian mixture filters employing EKFs and UKFs. These Gaussian mixture filters are denoted EGMF and UGMF in the following. Furthermore, a fully analytical Gaussian mixture filter (FAGMF) is employed as a baseline that performs component-wise moment matching for the prediction. All GMFs are initialized with different numbers of components, namely 1, 8, and 64 components. For each combination of noise level and number of components,

1000 simulation runs are performed, where each run consists of 50 time steps. The results are listed in Table 3.1.

The SAGMF clearly outperforms both EGMF and UGMF independent of the noise level or the used number of components. Furthermore, the estimation errors are almost identical with the baseline provided by the FAGMF. In terms of run-time, EGMF and AGMF are the fastest filters, but SAGMF is significantly faster than the UGMF, which suffers from the necessity of calculating high-dimensional matrix square roots. Furthermore, it requires a high number of on-line evaluations of the kinematics and measurement model, while in case of the SAGMF the number of functions evaluations is reduced significantly due to the small dimension of the sub-state $\boldsymbol{x}_s$.

It is important to point out that the outstanding performance of FAGMF is limited to a small number of applications. The SAGMF instead is of greater benefit as it is a general purpose filter, where both UGMF and FAGMF are its limiting cases.

3.4.2 Adaptive Gaussian Mixture Filter

The estimation error of Gaussian mixture filters significantly depends on the number of Gaussian components used. This number is typically defined by the user. The novel *adaptive Gaussian mixture filter* (AGMF) depicted in Figure 3.3 on the next page adapts the number of components dynamically and on-line and thus, directly tackles the refinement step of Algorithm 3. The nonlinear system and measurement models are linearized locally by means of statistical linear regression (see Section 2) at each component of the Gaussian mixture. The induced linearization error is quantified by means of the linearization error covariance matrix (2.28). Based on this error, a novel moment-preserving splitting procedure is proposed for introducing new mixture components. The component causing

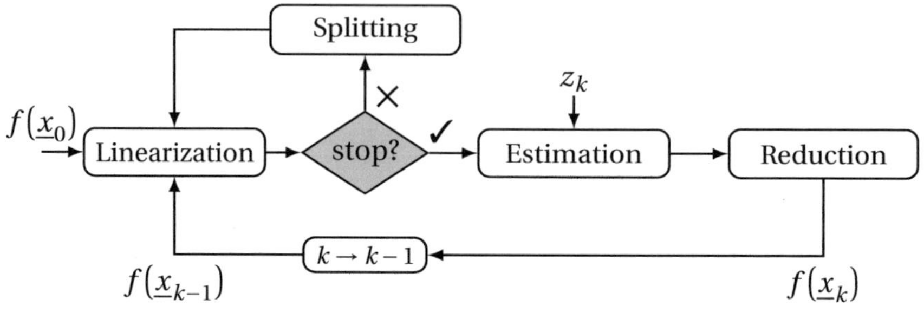

Figure 3.3: Flow chart of the adaptive Gaussian mixture filter.

the highest linearization error is selected, while splitting is performed in direction of the strongest nonlinearity, i.e., the strongest deviation between the nonlinear model and its linearized version. Both linearization and splitting are independent of the used statistical linear regression method, which makes the proposed filter versatilely applicable.

Component Selection

A straightforward way to select a Gaussian component for splitting is to consider the weights ω_i, $i = 1 \dots L$. The component with the highest weight is then split. This however does not take the nonlinearity of $\underline{g}(\cdot)$ in the support of the selected component into account. Since linearization is performed individually and locally, a more reasonable selection would be to consider also the induced linearization error of each component. For this purpose, statistical linear regression already provides an appropriate measure of the linearization error in form of the covariance matrix $\mathbf{C}_e$ in (2.28).

In order to easily assess the linearization error in the multi-dimensional case, the trace operator is applied to $\mathbf{C}_e$, which gives the measure

$$\varepsilon = \mathrm{Tr}(\mathbf{C}_e) \in [0, \infty) \,. \tag{3.19}$$

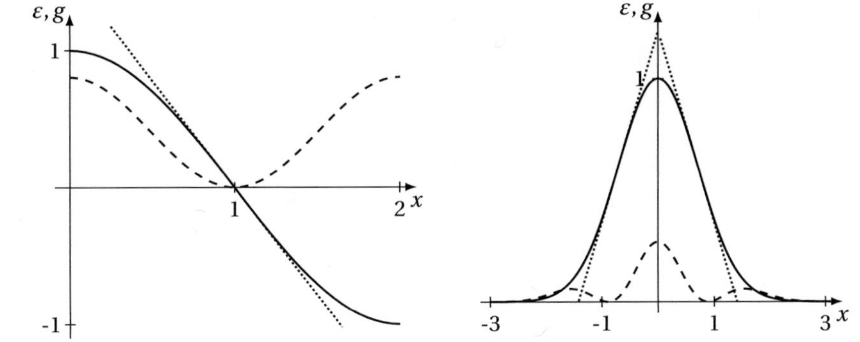

(a) $g(x) = \cos(x)$ **(b)** $g(x) = \exp\left(-x^2\right)$

Figure 3.4: Two examples of the linearization error quantification by means of the measures (3.19). The solid lines correspond to the function $g(.)$, the dashed lines to the measure (3.19), and the dotted lines indicate linearized versions of $g(.)$

Geometrically speaking, the trace is proportional to circumference of the covariance ellipsoid corresponding to $\mathbf{C}_e$. The larger $\mathbf{C}_e$ and thus the linearization error, the larger is ε. Conversely, the trace is zero, if and only if $\mathbf{C}_e$ is the zero matrix, i.e., $\varepsilon = 0 \Leftrightarrow \mathbf{C}_e = \mathbf{0}$. Hence, (3.19) is only zero, when there is no linearization error.

Example 13: Quantified Linearization Error

The error measure in (3.19) is applied to both scalar functions

$$g(\boldsymbol{x}) = \cos(\boldsymbol{x}) \quad \text{and} \quad g(\boldsymbol{x}) = \exp\left(-\boldsymbol{x}^2\right) ,$$

where $\boldsymbol{x} \sim \mathcal{N}(\mu, 1)$ has constant variance and the mean is moved along the x-axis. The corresponding linearization error values for both functions are depicted in Figure 3.4. The measure approaches zero whenever the function $g(.)$ is almost linear. This is the case for $\cos(x)$ if $x = 0$ and for $\exp\left(-x^2\right)$ if $x = \pm\sqrt{2}/2$ and $x \to \pm\infty$.

Besides the linearization error, the contribution of a component to the nonlinear transformation is important as well. That is, the probability mass of the component, which is given by its weight ω_i, has also to be taken into account. This avoids splitting of irrelevant components. Both ingredients are combined in the so-called *selection criterion*

$$i^* = \arg \max_{i=1\ldots L} \{s_i\} , \quad s_i \triangleq \omega_i^\gamma \cdot \big(1 - \exp(-\varepsilon_i)\big)^{1-\gamma} \tag{3.20}$$

Here, the term $1 - \exp(-\varepsilon_i)$ normalizes the linearization measure (3.19) to the interval $[0,1]$. For a geometric interpolation between weight and linearization error of component i, the parameter $\gamma \in [0,1]$ used. With $\gamma = 0$, selecting a component for splitting only focuses on the linearization error, while $\gamma = 1$ considers the weight only.

Splitting

Once a component causing a large linearization error is identified, new components are introduced by means of splitting the identified component. As mentioned above, splitting is an ill-posed problem due to the high number of free parameters. To reduce the degrees of freedom, splitting is performed given the following constraints:

1. *Along principal axes of a Gaussian:* This reduces the splitting problem of a multivariate Gaussian to the one-dimensional case. Furthermore, splitting becomes computationally cheap and numerically stable compared to arbitrary splitting directions.
2. *Splitting into two components:* Allows trading the reduction of the linearization error off against the increase of mixture components.
3. *Symmetric around the mean:* This minimizes the error introduced by splitting a Gaussian.
4. *Moment preserving:* The mean vector and covariance matrix of the split Gaussian and thus, of the entire mixture remains unchanged.

The restriction to split a Gaussian into two components is no limitation. As splitting is performed recursively by the AGMF within a time step, the newly introduced components can be split again in the next rounds if the linearization error is still too high.

Let $\omega \cdot \mathcal{N}(\underline{x}; \underline{\mu}, \mathbf{C})$ be the component considered for splitting. It is replaced by two components according to (3.13) with parameters

$$\begin{aligned} \omega_1 &= \omega_2 = \tfrac{\omega}{2} , \\ \underline{\mu}_1 &= \underline{\mu} + \sqrt{\lambda} \cdot \alpha \cdot \underline{v} , \quad \underline{\mu}_2 = \underline{\mu} - \sqrt{\lambda} \cdot \alpha \cdot \underline{v} , \\ \mathbf{C}_1 &= \mathbf{C}_2 = \mathbf{C} - \lambda \cdot \alpha \cdot \underline{v}\,\underline{v}^{\mathrm{T}} , \end{aligned} \tag{3.21}$$

where $\alpha \in [-1, 1]$ is a free parameter. The parametrization in (3.21) ensures moment preservation, i.e., the original Gaussian component and its split counterpart have the same mean and covariance. Furthermore, λ and $\underline{v}$ in (3.21) are a particular eigenvalue and eigenvector, respectively, of $\mathbf{C}$.

Splitting Direction

What remains an open question is the selection of an appropriate eigenvector for splitting. A straightforward choice might be the eigenvector with the largest eigenvalue as in [64, 86]. But since (3.20) determines the Gaussian component that causes the largest linearization error, merely splitting along the eigenvector with the largest eigenvalue does not take this error into account.

The key idea of the proposed criterion is to evaluate the deviation between the nonlinear transformation $\underline{g}(.)$ and its linearized version (2.26) along each eigenvector. The eigenvector with the largest deviation is then considered for splitting, i.e., the Gaussian is split in direction of the largest deviation in order to cover this direction with more Gaussians, which will reduce the error in subsequent linearization steps.

By means of the error term (2.27), the desired criterion for the splitting direction is defined as

$$d_l \triangleq \int_{\mathbb{R}} \underline{e}\big(\underline{x}_l(v)\big)^{\mathrm{T}} \cdot \underline{e}\big(\underline{x}_l(v)\big) \cdot \mathcal{N}\big(\underline{x}_l(v); \underline{\mu}, \mathbf{C}\big)\,\mathrm{d}v \tag{3.22}$$

with $\underline{x}_l(v) \triangleq \underline{\mu} + v \cdot \underline{v}_l$, $l = 1 \ldots n_x$, and $\underline{v}_l$ being the lth eigenvector $\mathbf{C}$. The integral in (3.22) cumulates the squared deviations along the lth eigenvector under the consideration of the probability at each point $\underline{x}_l(v)$. The eigenvector that maximizes (3.22) is then chosen for splitting. Unfortunately, due to the nonlinear transformation $\underline{g}(\cdot)$ this integral cannot be solved in closed-form in general. For an efficient and approximate solution, the sample point calculation schemes described in Section 2.2.5 are employed to approximate the Gaussian in (3.22) in direction of $\underline{v}_l$ by means of Dirac delta distributions. This automatically leads to a discretization of thc integral at a few but carefully chosen points.

Splitting Termination

As indicated in Figure 3.3, in every splitting round a *stopping criterion* is evaluated. Splitting stops, if at least one of the three following thresholds is reached:

Error threshold: The value s_i in the selection criterion (3.20) drops below $s_{\max} \in [0,1]$ for *every* component.

Component threshold: The number of mixture components excels $L_{\max}$.

Deviation threshold: The deviation between the original mixture $f(\underline{x})$ and the mixture obtained via splitting $\tilde{f}(\underline{x})$ excels $d_{\max} \in [0,1]$.

The deviation considered for the latter threshold is determined by means of the normalized version of the ISD according to

$$\bar{D}\big(f(\underline{x}), \tilde{f}(\underline{x})\big) \triangleq \frac{D\big(f(\underline{x}), \tilde{f}(\underline{x})\big)}{\int f(\underline{x})^2\,\mathrm{d}\underline{x} + \int \tilde{f}(\underline{x})^2\,\mathrm{d}\underline{x}} \in [0,1]\,. \tag{3.23}$$

Since splitting always introduces an approximation error to the original mixture, continuously monitoring the deviation limits this error.

Example 14: Shape Approximation

In order to demonstrate the effectiveness of splitting in direction of the strongest nonlinearity, the nonlinear growth process

$$\boldsymbol{y} = g(\underline{\boldsymbol{x}}) = \frac{\boldsymbol{\xi}}{2} + 5 \cdot \frac{\boldsymbol{\xi}}{1+\boldsymbol{\xi}^2} + \boldsymbol{w}$$

adapted from [102] is considered, where $\underline{\boldsymbol{x}} = [\boldsymbol{\xi}\ \boldsymbol{w}]^{\mathrm{T}} \sim \mathcal{N}\big([1,0]^{\mathrm{T}}, \mathbf{I}_2\big)$. To approximate the density of $\boldsymbol{y}$, the Gaussian $f(\underline{x})$ is split recursively into a Gaussian mixture, where the number of components is always doubled until a maximum of 64 components is reached. No mixture reduction and no thresholds $s_{\max}$, $d_{\max}$ are used. The true density of $\boldsymbol{y}$ is calculated via numerical integration.

Two different values for the parameter γ of the selection criterion (3.20) are used: $\gamma = 0.5$, which makes no preference between the component weight and the linearization error and $\gamma = 1$, which considers the weight only. Furthermore, a rather simple selection criterion is considered for comparison, where selecting a Gaussian for splitting is based on the weights only (as it is the case for $\gamma = 1$), while the splitting is performed in direction of the eigenvector with the largest eigenvalue.

Table 3.2 shows the KLD between the true density of $\boldsymbol{y}$ and the approximations obtained by splitting. The approximations of the proposed splitting scheme are significantly better than the approximations of the largest eigenvalue scheme. This follows from the fact that the proposed scheme not only considers the spread of a component. It also takes the linearization errors into account. In doing so, the Gaussians are always split along the eigenvector corresponding to $\boldsymbol{\xi}$, since this variable is transformed nonlinearly, while $\boldsymbol{w}$ is not. This is

Table 3.2: Approximation error (KLD × 10) for different splitting schemes and numbers of components.

splitting scheme	number of Gaussians						
	1	2	4	8	16	32	64
max. eigenvalue	2.01	0.77	0.64	0.47	0.39	0.21	0.26
$\gamma = 1$	2.01	0.77	0.59	0.34	0.20	0.12	0.07
$\gamma = 0.5$	2.01	0.77	0.40	0.22	0.07	0.03	0.02

different for the largest eigenvalue scheme, which wastes nearly half of the splits on $\boldsymbol{w}$.

The inferior approximation quality for $\gamma = 1$ compared to $\gamma = 0.5$ results from splitting components, which may have a high importance due to their weight but which do not cause severe linearization errors. Thus, splitting these components will not improve the approximation quality much.

In Figure 3.5, the approximate density of $\boldsymbol{y}$ is depicted for different numbers of mixture components for $\gamma = 0.5$. With an increasing number of components, the approximation approaches the true density very well.

3.4.3 Curvature-based Gaussian Mixture Reduction

Basically, the AGMF proposed above can operate with any Gaussian mixture reduction algorithm, but in order to maintain a high estimation performance, the employed reduction algorithm should provide a very good approximation of the original mixture. At the same time the reduction should be large enough in order to allow computationally feasible estimation. As mentioned in Section 3.3.2, global and pseudo-global reduction algorithms can provide an accurate approximation, but they

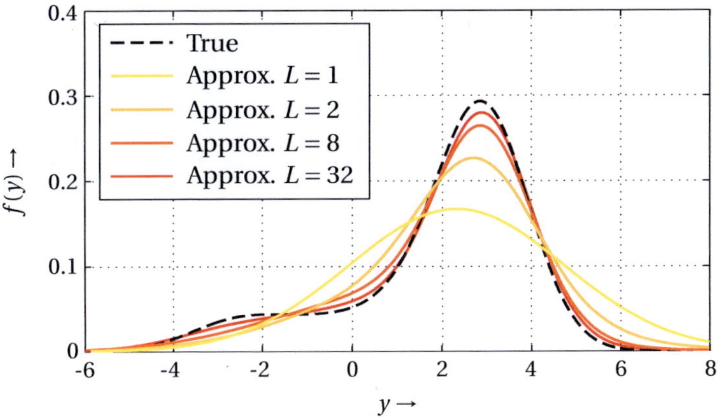

Figure 3.5: True density function of $\boldsymbol{y}$ (black, dashed) and approximations with an increasing number of mixture components for $\gamma = 0.5$.

suffer from a high complexity. This drawback becomes even more severe, when the number of mixture components is initially very large. This situation, however, appears quite often in filtering applications, especially when Gaussian mixtures are multiplied or when splitting adds many components due to a high state dimension.

Especially in cases, where the number of components is large, it is possible to find a reduced mixture merely by setting the weights ω_i of many components to zero and still, the reduced mixture approximates the original mixture sufficiently well.

Example 15: Weight Optimization

Consider a Gaussian mixture consisting of 20 components as depicted in Figure 3.6a. Intuitively one would guess that two to three Gaussians should be sufficient to provide an approximation of the original mixture that captures both modes. The remaining components seem to be redundant. In Figure 3.6b the extreme case is depicted, where the weights of all components except of two are set

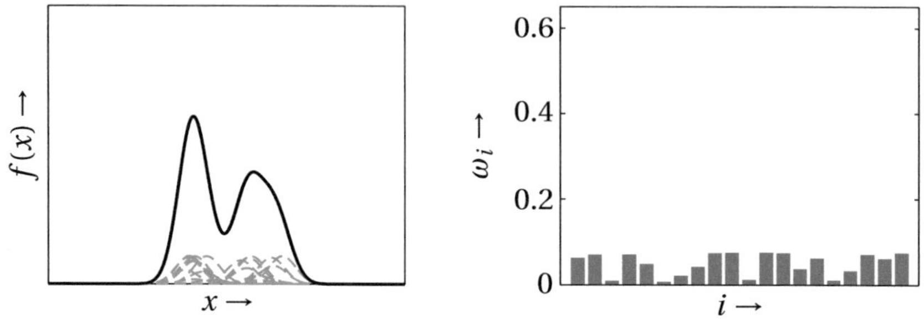

(a) Original Gaussian mixture consisting of 20 components.

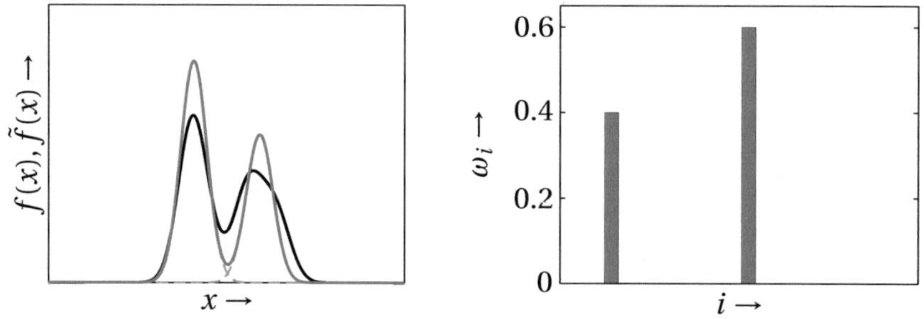

(b) Reduced Gaussian mixture consisting of two components.

Figure 3.6: A Gaussian mixture and its reduced version. Left: original Gaussian mixture (black), reduced mixture (dark gray), and the mixture components (light gray, dashed). Right: the mixture weights.

to zero. The weights of the remaining two components are merely normalized such that their sum is equal to one. The resulting reduced mixture captures both modes. However, there is still room for improvement, e.g., by optimizing the weight or by allowing one more component.

The *superficial Gaussian mixture reduction* (SGMR) introduced next is a global reduction algorithm that is based on minimizing the curvature—

maximizing the smoothness—of the reduced mixture while keeping the ISD low. By means of using the curvature it is possible to identify similar components globally and to remove these components from the density. Carrying the idea that similar components may be dropped a step further, the trade-off between curvature and approximation error is minimized by merely optimizing the weights, i.e., assigning zero weights to reduced components. This approach is computationally feasible as no other parameters of the mixture need to be optimized. It further allows a simple and efficient implementation based on standard quadratic program (QP) solvers. Additionally, this weight-only optimization alleviates the model selection problem as the final number of components is automatically derived from setting the trade-off between error and roughness. This hyperparameter may be automatically optimized as well. In cases where the resulting number of components still is too large from a computational perspective, existing global reduction algorithm can be employed ex post. Thanks to the already reduced mixture, the computational load of the global algorithm can be lowered significantly.

Quadratic Program

By focusing on the weights only, the mixture reduction problem can be formulated as a quadratic program

$$\begin{aligned} \min_{\underline{\tilde{\omega}}} \quad & \underline{\tilde{\omega}}^{\mathrm{T}} \mathbf{Q} \, \underline{\tilde{\omega}} - \underline{q}^{\mathrm{T}} \underline{\tilde{\omega}} \\ \text{s.t.} \quad & \underline{1}^{\mathrm{T}} \cdot \underline{\tilde{\omega}} = 1 \, , \\ & \underline{0} \preceq \underline{\tilde{\omega}} \, , \\ & \underline{0} = \textstyle\sum_{i=1}^{L} \underline{\mu}_i \cdot (\omega_i - \tilde{\omega}_i) \, , \end{aligned} \tag{3.24}$$

where $\underline{\tilde{\omega}}^{\mathrm{T}} \triangleq [\tilde{\omega}_1 \, \ldots \, \tilde{\omega}_L]$ is the vector of all weights of $\tilde{f}(\underline{x})$ to be optimized. The symmetric matrix $\mathbf{Q} \triangleq \mathbf{D} + \lambda \mathbf{R}$ comprises the positive semi-definite matrices $\mathbf{D}$ and $\mathbf{R}$, where $\mathbf{D}$ originates from the ISD and $\mathbf{R}$ is a *roughness penalty* measuring the curvature of $\tilde{f}$. The vector $\underline{q} \triangleq 2\,\underline{d}$, where $\underline{d}^{\mathrm{T}} \triangleq \underline{\tilde{\omega}}^{\mathrm{T}} \mathbf{D}$

depends on the ISD as well. The hyperparameter λ governs the trade-off between $\mathbf{D}$ and $\mathbf{R}$. It needs to be determined by means of generic model selection algorithms, see e.g. [164]. For small values of λ, the ISD will be weighted relatively higher than the curvature. This results in more components in the reduced mixture $\tilde{f}$ and less approximation error. For large values of λ, the curvature will be weighted higher enforcing more reduction and approximation error.

The constraints in the QP assert the integration of the probability mass to one, the positivity of the density, and that $\tilde{f}$ and f have identical means.

It is important to note the number of components in f and $\tilde{f}$ is identical. After solving (3.24), many weights will be zero or close to zero. Thus, the corresponding components no longer contribute to the mixture and can be discarded.

The components of the ISD in (3.24) are obtained, when examining (3.11) for all parameters but the weights of $\tilde{f}$ as fixed. It turns out that (3.11) merely consists of linear and quadratic terms of the weights $\tilde{\omega}_i$. Thus, the ISD can be written as

$$D\big(f(\underline{x}),\tilde{f}(\underline{x})\big) = \underline{\tilde{\omega}}^{\mathrm{T}}\mathbf{D}\,\underline{\tilde{\omega}} - 2\underline{d}^{\mathrm{T}}\underline{\tilde{\omega}} + c\,, \tag{3.25}$$

where the matrix $\mathbf{D}$ corresponds to the self-similarity of $\tilde{f}$, the vector $\underline{d}$ encodes the cross-similarity between f and $\tilde{f}$, and the constant $c \triangleq \underline{\omega}^{\mathrm{T}}\mathbf{D}\,\underline{\omega}$ corresponds to the self-similarity of f.

The roughness of $\tilde{f}$ is interpreted as the curvature κ of the density's surface. Since the curvature (see e.g. [36]) is signed and a function of the position on the surface, a quantification in terms of the integral squared curvature is sought. The key idea is to derive an (approximate) upper bound of the squared curvature of the mixture. The derivation is based on the point-wise squared curvature $\kappa(\underline{x})$ for a density $\tilde{f}$, for which an upper bound $\breve{\kappa}(\underline{x})$ is determined, integrated over the entire domain of $\underline{x}$, i.e., $R\big(\tilde{f}(\underline{x})\big) = \int \breve{\kappa}(\underline{x})^2 \, \mathrm{d}\underline{x}$. For the 1D and 2D case, the upper bounds are

$$\breve{\kappa}(x)^2 \triangleq \left(\underline{\tilde{\omega}}^{\mathrm{T}} \underline{\tilde{f}}_{xx}\right)^2 , \quad \breve{\kappa}\left(\underline{x}\right)^2 \triangleq \left(\underline{\tilde{\omega}}^{\mathrm{T}} \underline{\tilde{f}}_{xx} - 2\,\underline{\tilde{\omega}}^{\mathrm{T}} \underline{\tilde{f}}_{x} \underline{\tilde{\omega}}^{\mathrm{T}} \underline{\tilde{f}}_{y} \underline{\tilde{\omega}}^{\mathrm{T}} \underline{\tilde{f}}_{xy} + \underline{\tilde{\omega}}^{\mathrm{T}} \underline{\tilde{f}}_{yy}\right)^2 ,$$

where $\underline{\tilde{f}}_{x}$ and $\underline{\tilde{f}}_{xy}$ being the first-order and second-order derivatives of $\underline{\tilde{f}}$, with $\underline{\tilde{f}}$ defined as in (3.2). These upper bounds of the integral squared mean curvature are obtained by dropping the denominator and positive summands.

For the weight optimization, the upper bound of the curvature is formulated as a quadratic form $\underline{\tilde{\omega}}^{\mathrm{T}}\, \mathbf{R}\, \underline{\tilde{\omega}}$. The elements of $\mathbf{R}$ may be obtained as follows. For the 1D case the elements of $\mathbf{R}$ are given by $\mathbf{R}_{ij} = \int_{\mathbb{R}} \tilde{f}^{(i)}_{xx} \, f^{(j)}_{xx} \, \mathrm{d}x$, where $\tilde{f}^{(i)}$ refers to the ith mixture component of $\tilde{f}$. For the 2D case, the following approximation is used

$$\begin{aligned}\left(\underline{\tilde{\omega}}^{\mathrm{T}} \underline{\tilde{f}}_{xx} - 2\,\underline{\tilde{\omega}}^{\mathrm{T}} \underline{\tilde{f}}_{x} \underline{\tilde{\omega}}^{\mathrm{T}} \underline{\tilde{f}}_{y} \underline{\tilde{\omega}}^{\mathrm{T}} \underline{\tilde{f}}_{xy} + \underline{\tilde{\omega}}^{\mathrm{T}} \underline{\tilde{f}}_{yy}\right)^2 \\ \approx \left(\underline{\tilde{\omega}}^{\mathrm{T}} \left[\underline{\tilde{f}}_{xx} - 2\,\underline{\tilde{f}}_{x} \underline{\tilde{f}}_{y}^{\mathrm{T}} \underline{\tilde{f}}_{xy} + \underline{\tilde{f}}_{yy}\right]\right)^2 ,\end{aligned}$$

which leads to the elements

$$\mathbf{R}_{ij} = \int_{\mathbb{R}^2} \left(\tilde{f}^{(i)}_{xx} - 2\,\tilde{f}^{(i)}_{x} \, \tilde{f}^{(i)}_{y} \, \tilde{f}^{(i)}_{xy} + \tilde{f}^{(i)}_{yy}\right) \cdot \left(\tilde{f}^{(j)}_{xx} - 2\,\tilde{f}^{(j)}_{x} \, \tilde{f}^{(j)}_{y} \, \tilde{f}^{(j)}_{xy} + \tilde{f}^{(j)}_{yy}\right) \mathrm{d}\underline{x} \, . \tag{3.26}$$

For the 1D curvature, the $\mathbf{R}_{ij}$ may be calculated in closed form. Note for a 2D density the curvature is not unique, as it is calculated from the minimum and maximum curvature in the principal directions at each point $\underline{x}$, which may be multiplied (Gaussian curvature) or averaged (mean curvature) [36]. For arbitrary Gaussian mixture densities, the terms in (3.26) may only be calculated numerically or need to be approximated further.

Algorithm

The entire SGMR algorithm comprises three parts: the *pre-processing* of the components of the quadratic forms and the hyperparameter, the weight optimization by the solution of (3.24), and a fast *post-optimization* of the already reduced set of weights from (3.24).

The *pre-processing* consists of calculating the matrix **D** and vector $\underline{d}$ corresponding to the ISD as well as the matrix **R** describing the curvature of the mixture's surface.

As the matrix **Q** is positive semi-definite, the QP is convex and can be solved efficiently and globally optimal by any standard solver (see Appendix C). The resulting weights are compared against a threshold $\varepsilon \ll 1$. This leads to reduced weights $\tilde{\omega}_i^+ \geq \varepsilon$, $i = 1 \ldots M \ll L$.

The purpose of the *post-optimization* is an adaptation of the already reduced weights $\underline{\tilde{\omega}}^+$, aimed at improving the accuracy, by neglecting the curvature and only minimizing the ISD with respect to $\underline{\tilde{\omega}}^+$. The corresponding QP is similar to (3.24) but without the curvature matrix **R**.

Example 16: 1D Reduction

The proposed SGMR is compared against the following reduction algorithms: *Pruning* [20] of all but the components with the highest weights, *West*'s merging [208], *Salmond*'s clustering algorithm [158], *Williams*' merging algorithm [211], *Runnalls*' algorithm [156], and *PGMR* [90]. For SGMR, two variants are considered, one with post-optimization and one without post-optimization.

For evaluation, univariate Gaussian mixtures with a number of components $L \in \{40, 80, 120, 160, 200\}$ are used. The mixture parameters are drawn uniformly at random from the intervals $\tilde{\alpha} \in [0.05, 0.5]$, $\tilde{\mu} \in [0, 3]$, and $\tilde{\sigma} \in [0.09, 0.5]$. For each number of components L, 50 Monte Carlo simulation runs are performed. For SGMR, the hyperparameter λ is set to 500 and the deletion threshold ε is $1e^{-4}$. The maximum error threshold of PGMR is set to 1%.

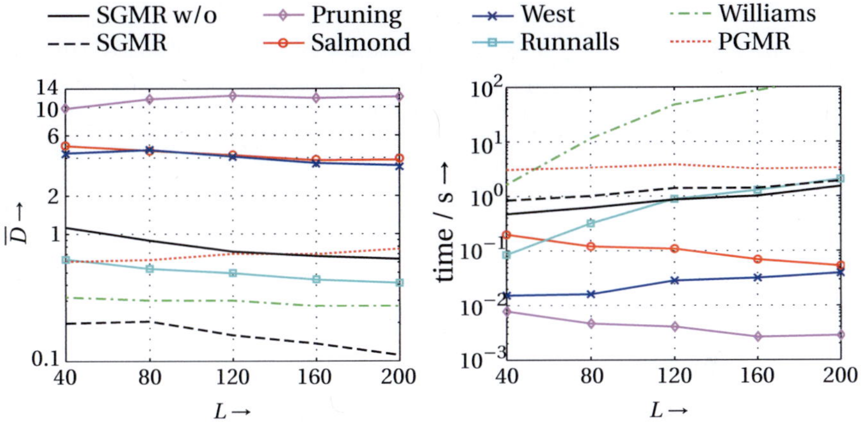

Figure 3.7: Reduction error (left) and runtime (right) of different reduction algorithms for increasing number of components. The results are averages over 50 Monte Carlo runs. The average reduction error is multiplied by 100 for better readability.

Pruning, West, Salmond, Williams, and Runnalls require a user-defined threshold on the number of components to which the given Gaussian mixture has to be reduced. Since SGMR reduces a Gaussian mixture in a completely different fashion and thus, to ensure a fair comparison, the number of components resulting from SGMR with post-optimization is used as threshold for these approaches. In order to quantify the reduction error, the normalized ISD (3.23) is used.

In Figure 3.7, the average reduction errors and the average computation times for all L are shown. It can be seen that SGMR provides the lowest reduction error. Closest to SGMR is Williams' algorithm, but this algorithm clearly suffers from its high computational demand. Salmond's and West's methods perform similarly. Both are very fast, but their approximation quality is the worst except of pruning. In terms of the reduction error, the results of Runnalls' method are in between of SGMR with and without post-optimization. But for an

Table 3.3: Number of components in the reduced Gaussian mixture for different reduction algorithms. The results are averages over 50 MC runs.

L	SGMR w/o	PGMR	SGMR all others
40	21.66	7.34	20.54
80	30.88	7.42	29.5
120	37.18	7.58	35.86
160	42.86	6.78	41.8
200	45.68	6.7	44.98

increasing number of components L in the original mixture it becomes computationally more expensive than both SGMR methods. Overall, SGMR provides the best trade-off between reduction error and computation time.

The reduction performance of SGMR improves with a larger number of components in the original mixture. As listed in Table 3.3, SGMR reduces to about 50% if the number of components of the original mixture is $L = 40$, while for $L = 200$ only 22% of the components remain. Since SGMR merely adapts the weights $\underline{\tilde{\omega}}$, a larger number of components is advantageous for SGMR for a better exploitation of redundancies. This leads to a stronger reduction by a simultaneously lower reduction error. Furthermore, the comparison between SGMR and SGMR without post-optimization shows that the post-optimization always lowers both the reduction error and the number of components.

For 1D mixtures, PGMR clearly is the best reduction algorithm. The reduction error is close to SGMR without post-optimization, but the number of components in the reduced mixture is significantly lower (see Table 3.3). However, a straightforward extension to multivariate mixtures is not possible as only axis-aligned Gaussian components can be utilized for representing the reduced mixture.

3.5 Summary

Three algorithms supporting the generic Gaussian mixture filtering as sketched in Algorithm 3 have been proposed in this chapter, namely semi-analytic Gaussian mixture filter (SAGMF), adaptive Gaussian mixture filter (AGMF), and superficial Gaussian mixture reduction (SGMR). These algorithms aggregate following contributions:

- *Component-wise Rao-Blackwellization for nonlinear-nonlinear decomposition:* The decomposition of conditionally integrable functions already proposed for Gaussian filters can be applied straightforwardly in a mixture setup.
- *General linearization error measure for LRKFs:* The proposed linearization error measure exploits the error covariance that is provided as a side product of statistical linear regression. Thus, the computational overhead for quantifying the linearization error is low and the measure can be utilized by any LRKF.
- *Splitting along the strongest nonlinearity:* New components are introduced at the strongest nonlinearity in order to reduce linearization errors most effectively.
- *Scalable and moment-preserving splitting:* Splitting is performed carefully as components are only added where they are really required. This facilitates scaling with high state dimensions. In addition the mean and covariance of the Gaussian mixture are preserved.
- *Mixture reduction via weight-optimization:* To lower the computational demand of global mixture reduction but at the same to benefit from a low reduction errors, components are removed by considering their weights only. For this purpose, a novel QP compromising reduction error and roughness has been proposed.

As with the contributions made for Gaussian filtering, the above contributions and algorithms can and should be used in combination. The nonlinear sub-state that cannot be treated analytically by the SAGMF, can be processed by means of the AGMF, where the linearization error can be kept at a minimum. The newly introduced components due to splitting have to be reduced from time to time in order to keep the computational demand bounded. For this purpose, the SGMR can be employed.

4

Gaussian Process Filtering

So far it was assumed that the nonlinear relation between the current state $\underline{\boldsymbol{x}}_k$ and the next state $\underline{\boldsymbol{x}}_{k+1}$ or the measurement $\underline{\boldsymbol{z}}_k$ are known. In many applications, however, this assumption is no longer valid. Instead of a functional representation of these relations, only labeled data sets are given. Hence, at first the functional representation has to be learned from the data before the actual filtering can be performed. In this chapter, *Gaussian process* (GP) regression for learning the nonlinear model from data is introduced[1]. A GP defines a Gaussian prior density over functions, which can be turned into a posterior density over functions using Bayesian inference when data is present. Evaluating the resulting posterior model at a finite number of inputs yields a Gaussian distribution of functions values.

1 In geostatistics GP regression is known as *kriging* [57].

A GP model is *non-parametric*[2], which has the benefit that one has not to worry about selecting the wrong model to fit the data. Key element of a GP is the so-called *covariance function* or kernel function, which specifies the correlation between the data points. In contrast to other kernel methods like support vector machines (SVMs, [165]) or kernel least squares [62] a GP in addition provides confidence intervals thanks to the density representation of the regression result. Section 4.2 gives a short introduction to the most commonly used covariance functions and how the parameters of these functions can be learned from data. In Section 4.3, the major drawback of the non-parametric nature of GPs is addressed: the high computational demand given large-scale data sets. The state-of-the-art of Bayesian filtering given GP models is surveyed in Section 4.4. The contributions to GP filtering and efficient learning made in Papers H–K are summarized in Section 4.5.

4.1 Gaussian Processes

For GP regression, it is assumed that a set $\mathcal{D} = \{(\underline{x}_1, y_1), \ldots, (\underline{x}_n, y_n)\}$ of *training* data is drawn from the nonlinear mapping

$$y = g(\underline{x}) + \boldsymbol{w}\,, \quad \text{with } \boldsymbol{w} \sim \mathcal{N}(0, \sigma^2) \tag{4.1}$$

where $\underline{x}_i \in \mathbb{R}^{n_x}$ are the inputs and $y_i \in \mathbb{R}$ are the observations or outputs, for $i = 1 \ldots n$. For brevity reasons, $\mathbf{X}_{\mathcal{D}} \triangleq [\underline{x}_1 \ \ldots \ \underline{x}_n]$ are all inputs and $\underline{y}_{\mathcal{D}}^{\mathrm{T}} \triangleq [y_1 \ \ldots \ y_n]$ are the corresponding observations in the following.

A GP is used to infer the latent function $g(.)$ from the data $\mathcal{D}$. It is completely defined by a *mean function* $\mu(\underline{x}) \triangleq \mathrm{E}\{\mathbf{g}(\underline{x})\}$, which specifies the

2 The term non-parameteric is a bit misleading. Non-parametric methods like GPs still have parameters, but in contrast to parametric approaches, the learned model has no characteristic structure and parameters. The structure/appearance of the model is derived from training data [32].

expected output value, and a positive semi-definite *covariance function* $\kappa(\underline{x},\underline{x}') \triangleq \operatorname{cov}\{\mathbf{g}(\underline{x}), \mathbf{g}(\underline{x}')\}$, which specifies the covariance between pairs of inputs. In the following, the zero mean function $\mu(\underline{x}) = 0$ is considered, which is a reasonable assumption and not a limitation in general [54, 139]. For a detailed discussion on covariance functions see Section 4.2.

A GP forms a Gaussian distribution over functions and thus, one can write $\mathbf{g} \sim \mathcal{GP}\left(\mu(\underline{x}), \kappa(\underline{x},\underline{x}')\right)$. Accordingly, for a finite data set $\mathcal{D}$ the resulting density function of the outputs is a multivariate Gaussian

$$f(\underline{g}|\mathbf{X}_{\mathcal{D}}) = \mathcal{N}(\underline{0}, \mathbf{K}) \tag{4.2}$$

with $\underline{g}^{\mathrm{T}} \triangleq \left[g(\underline{x}_1) \; \dots \; g(\underline{x}_n)\right]$ being the vector of latent function values at the training data inputs. This density is the so-called GP prior with *kernel* matrix $\mathbf{K} \triangleq \kappa(\mathbf{X}_{\mathcal{D}}, \mathbf{X}_{\mathcal{D}})$ comprising the elements $(\mathbf{K})_{ij} = \kappa(\underline{x}_i, \underline{x}_j), \forall \underline{x}_i, \underline{x}_j \in \mathcal{D}$. In Figure 4.1a on the next page an exemplary GP prior is depicted.

The *posterior* density of $g(.)$ for an arbitrary input $\underline{x}$ results from marginalizing the latent function over the training set according to

$$f(g(\underline{x}) \mid \underline{x}, \mathcal{D}) = \int f\left(g(\underline{x}), \underline{g} \,\middle|\, \underline{x}, \mathcal{D}\right) \mathrm{d}\underline{g} \,. \tag{4.3}$$

The integrand in (4.3) results from solving the Bayesian inference

$$f\left(g(\underline{x}), \underline{g} \,\middle|\, \underline{x}, \mathcal{D}\right) = c \cdot f\left(\underline{y}_{\mathcal{D}} \,\middle|\, \underline{g}\right) \cdot f\left(g(\underline{x}), \underline{g} \,\middle|\, \underline{x}, \mathbf{X}_{\mathcal{D}}\right) , \tag{4.4}$$

where $f\left(g(\underline{x}), \underline{g} \,\middle|\, \underline{x}, \mathbf{X}_{\mathcal{D}}\right)$ is the GP prior as in (4.2), but extended with the input $\underline{x}$. The likelihood $f\left(\underline{y}_{\mathcal{D}} \,\middle|\, \underline{g}\right)$ is given by

$$f\left(\underline{y}_{\mathcal{D}} \,\middle|\, \underline{g}\right) = \prod_{i=1}^{n} \mathcal{N}\left(y_i; g(\underline{x}_i), \sigma^2\right) \tag{4.5}$$

according to the model (4.1). Hence, all involved density functions are Gaussian and thus, the inference in (4.4) as well as the marginalization in

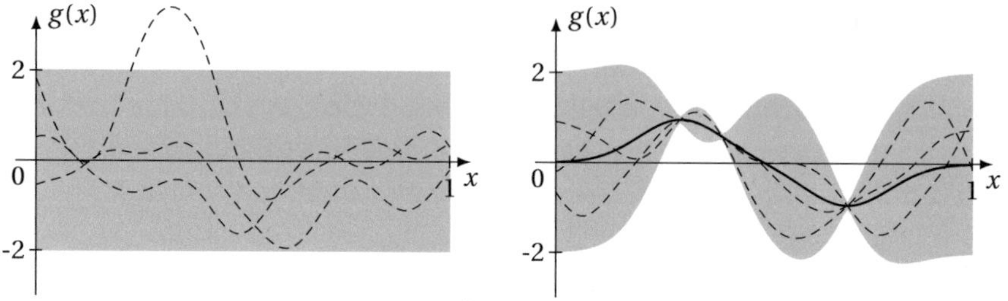

(a) Three sample functions drawn from a GP prior.

(b) The posterior GP after observing three data points. The solid line corresponds to the posterior mean (4.6).

Figure 4.1: Gaussian process prior and posterior. The gray areas indicate 2-sigma confidence regions.

(4.3) can be solved analytically. The desired posterior density in (4.3) is Gaussian with mean and variance

$$\mu_g(\underline{x}) = \mathrm{E}\{\mathbf{g}(\underline{x})\} = \underline{k}^{\mathrm{T}} \cdot \mathbf{K}_y^{-1} \cdot \underline{y}_{\mathcal{D}} , \tag{4.6}$$

$$\sigma_g^2(\underline{x}) = \mathrm{var}\{\mathbf{g}(\underline{x})\} = \kappa(\underline{x}, \underline{x}) - \underline{k}^{\mathrm{T}} \cdot \mathbf{K}_y^{-1} \cdot \underline{k} , \tag{4.7}$$

respectively, with $\underline{k}^{\mathrm{T}} \triangleq \kappa(\underline{x}, \mathbf{X}_{\mathcal{D}}) = [\kappa(\underline{x}_1, \underline{x}) \ldots \kappa(\underline{x}_n, \underline{x})]$ and $\mathbf{K}_y \triangleq \mathbf{K} + \sigma^2 \cdot \mathbf{I}_n$.

Example 17: GP Posterior

The latent function is chosen to be $g(x) = \sin(2\pi x)$ and the noise variance is $\sigma^2 = 0.02^2$. As covariance function the squared exponential function in (4.8) on page 128 is considered with hyperparameters $\alpha = 1$ and $\lambda = 0.1$. The training data comprises the inputs $\mathbf{X}_{\mathcal{D}} = [0.3\ 0.4\ 0.7]$ and outputs $\underline{y}_{\mathcal{D}}^{\mathrm{T}} = [0.96\ 0.59\ -0.93]$. The corresponding posterior GP is depicted in Figure 4.1b. To obtain this plot, the GP is evaluated for 100 values of $x \in [0\ 1]$, where the 100 values are chosen to be equidistant.

It can be seen that near a data point the posterior mean (4.6) is almost identical with the latent function and the posterior variance (4.7) is reduced significantly. Appart from data, the mean goes back to zero and the variance grows towards the value of the prior variance in Figure 4.1a. As a GP defines a distribution over functions, there is an infinite number of functions that can explain the data. Three of them are depicted in Figure 4.1b.

Based on (4.3), it is also possible to determine the density $f(y|x,\mathcal{D})$ of the output y. This density is also Gaussian with mean as in (4.6) and with variance

$$\sigma_y^2 = \sigma_g^2 + \sigma^2 .$$

If instead of the zero mean function an arbitrary mean function $\mu(\underline{x})$ is used, the posterior mean (4.6) changes to

$$\mu_g(\underline{x}) = \mu(\underline{x}) + \underline{k}^{\mathrm{T}} \cdot \mathbf{K}_y^{-1} \cdot (\underline{y}_{\mathcal{D}} - \underline{m})$$

with $\underline{m}^{\mathrm{T}} \triangleq [\mu(\underline{x}_1) \ \ldots \ \mu(\underline{x}_n)]$, while the posterior variance remains the same. For further reading on GP regression please be referred to [144].

4.2 Covariance Functions

So far it was assumed, that the covariance function is known. In real-world applications however, this assumption is typically not valid. Selecting an appropriate covariance function and determining its parameters is a crucial task. For a GP, the covariance function is of same importance as the kernel function in a SVM. It determines the relation between data points, characterizes the smoothness of the (latent) function, and specifies the impact of a data point on the overall function.

Not every function mapping two vectors to the real line can be considered as a covariance function. It has to be positive semi-definite and symmetric, i.e., the resulting kernel matrix has to be positive semi-definite[3] and symmetric matrix. This constraint is important as the kernel matrix acts as the covariance matrix of the Gaussian prior density (4.2).

4.2.1 Examples

In the following, some examplary covariance functions used throughout this thesis are introduced. Other commonly employed functions are discussed in [144].

Stationary Covariance Functions

Basically, covariance functions can be separated in two classes: stationary and non-stationary functions. A covariance function is called *stationary*[4] if it can be written as a function of the deviation $\Delta\underline{x} \triangleq \underline{x} - \underline{x}'$, i.e., $\kappa(\underline{x}, \underline{x}') = \kappa(\Delta\underline{x})$, for two arbitrary inputs $\underline{x}$ and $\underline{x}'$.

One of the most widely used covariance functions for GPs, but also in many other kernel-based machine learning algorithms like SVMs or radial-basis networks, is the *squared exponential* (SE)

$$\kappa(\Delta\underline{x}) = \alpha^2 \cdot \exp\left(-\tfrac{1}{2} \cdot \Delta\underline{x}^{\mathrm{T}} \Lambda^{-1} \Delta\underline{x}\right) . \tag{4.8}$$

Here, $\Lambda = \operatorname{diag}\left(\left[\lambda_1 \ \ldots \lambda_{n_x}\right]\right)$ is a diagonal matrix of the *characteristic length-scales* λ_i for each input dimension i and α^2 is the variance of the latent function $g(.)$. Such parameters of the covariance function are called the *hyperparameters* of the GP.

3 All eigenvalues of the matrix are non-negative.

4 The term stationary stems from stochastic process theory, where a process is called (weakly) stationary if it possesses a constant mean function and a translation-invariant covariance function [100].

Non-Stationary Covariance Functions

Covariance functions that directly depend on the two inputs $\underline{x}$ and $\underline{x}'$ are called *non-stationary*. A widely used non-stationary kernel is the *polynomial* covariance function

$$\kappa(\underline{x}, \underline{x}') = \alpha^2 \cdot \left(\underline{x}^{\mathrm{T}} \cdot \underline{x}' + c\right)^p , \tag{4.9}$$

with degree $p \in \mathbb{N}$, bias c, and variance α^2. Another popular covariance function is the *neural network* (NN) kernel

$$\kappa(\underline{x}, \underline{x}') = \alpha^2 \cdot \sin^{-1}\left(\frac{\underline{x}^{\mathrm{T}} \Lambda^{-2} \underline{x}'}{\sqrt{\phi(\underline{x})\phi(\underline{x}')}}\right) , \tag{4.10}$$

with $\phi(\underline{x}) \triangleq 1 + \underline{x}^{\mathrm{T}} \Lambda^{-2} \underline{x}$, variance α^2, and the diagonal matrix Λ of characteristic length-scales.

Constructing Covariance Functions

Covariance functions can be constructed from existing ones like the above mentioned. Let $\kappa_1(\underline{x}, \underline{x}')$ and $\kappa_2(\underline{x}, \underline{x}')$ be two covariance functions, the

- *Sum* of both covariance functions $\kappa_1(\underline{x}, \underline{x}') + \kappa_2(\underline{x}, \underline{x}')$,
- *Product* of both covariance functions $\kappa_1(\underline{x}, \underline{x}') \cdot \kappa_2(\underline{x}, \underline{x}')$, and
- *Scaling* $\phi(\underline{x}) \cdot \kappa_1(\underline{x}, \underline{x}') \cdot \phi(\underline{x}')$ with a deterministic function $\phi(\underline{x})$

lead again to valid covariance functions.

Example 18: SE plus Noise

In many applications, the SE kernel (4.8) is augmented by the noise variance of the model (4.1) according to

$$\kappa(\Delta\underline{x}) = \alpha^2 \cdot \exp\left(-\tfrac{1}{2} \cdot \Delta\underline{x}^{\mathrm{T}} \Lambda^{-1} \Delta\underline{x}\right) + \sigma^2 \cdot \delta(\Delta\underline{x}) . \tag{4.11}$$

This allows treating the noise variance σ^2 as an additional hyperparameter to be learned from data (see next section).

4.2.2 Hyperparameter Learning

Each covariance function possesses several parameters, e.g., the variance α^2 and the length-scale matrix Λ in case of the SE function (4.8). As mentioned above, these so-called hyperparameters need to be adjusted such that the covariance function fits to the given application[5]. In case of GPs, the hyperparameters can be determined or *learned* from the given training data $\mathcal{D}$. In the following, $\underline{\theta}$ comprises all hyperparameters of a given covariance function.

Example 19: Varying Hyperparameters

The latent function is chosen to be $g(x) = \sin(x)$. 20 data points are drawn from this model. Based on this training data, three different GPs with covariance function (4.11) are learned. The resulting GPs are depicted in Figure 4.2. The hyperparameters chosen for the GP in Figure 4.2a are close to optimal. Accordingly, the GP approximates the latent function very well and the posterior variances are adequate. In Figure 4.2b and Figure 4.2c, however, the hyperparameters are not chosen appropriately. While the GP in Figure 4.2b is clearly overfitting, the GP in Figure 4.2c provides an oversmoothed result.

5 Hyperparameter learning is part of the *model selection* problem (recall Sections 3.3.2 and 3.4.3).

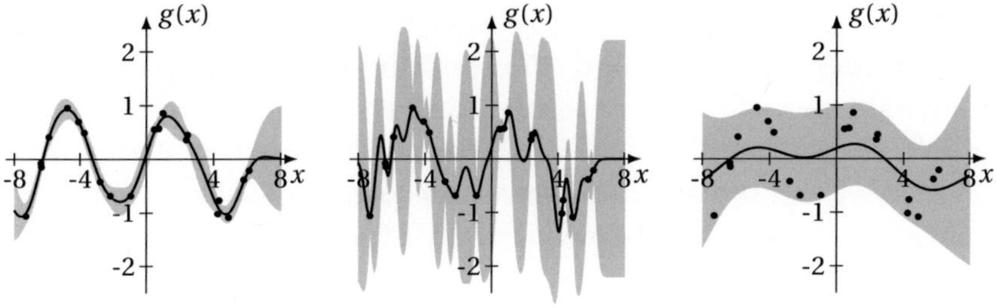

(a) $\underline{\theta}^{\mathrm{T}} = [1\ 0.5\ 0.1]$ **(b)** $\underline{\theta}^{\mathrm{T}} = [0.3\ 1.1\ 0.01]$ **(c)** $\underline{\theta}^{\mathrm{T}} = [3\ 1.2\ 0.9]$

Figure 4.2: Posterior GPs for different hyperparameters $\underline{\theta}^{\mathrm{T}} = [\alpha\ \lambda\ \sigma]$. The data points are indicated by the black circles.

To determine appropriate hyperparameters, the standard approach proposed in [118] is *evidence maximization*, which essentially tries to maximize the (positive) log-likelihood (see Appendix B.4)

$$\begin{aligned}\underline{\theta}^* &= \arg\max_{\underline{\theta}} \left\{ \log f\left(\underline{y}_{\mathcal{D}} \middle| \underline{g}, \underline{\theta}\right) \right\} \\ &= \arg\max_{\underline{\theta}} \left\{ \underbrace{-\tfrac{1}{2} \cdot \underline{y}_{\mathcal{D}}^{\mathrm{T}} \mathbf{K}_y^{-1} \underline{y}_{\mathcal{D}}}_{\text{data-fit}} \underbrace{- \tfrac{1}{2} \log |2\pi \mathbf{K}_y|}_{\text{complexity}} \right\} ,\end{aligned} \tag{4.12}$$

where the likelihood $f\left(\underline{y}_{\mathcal{D}} \middle| \underline{g}, \underline{\theta}\right)$ corresponds to (4.5) and $\mathbf{K}_y$ depends on $\underline{\theta}$. The maximization in (4.12) yields a trade-off between fitting the data as good as possible and obtaining a model of low complexity. Hence, this maximization is a realization of the famous *Occam's razor*[6], which aims for the simplest model possible that is consistent with the data.

Evidence maximization is a non-convex optimization problem, for which no closed-form solution exists and which highly depends on the given

6 "Pluralitas non est ponenda sine necessitate" (plurality should not be posited without necessity) stated by William of Ockham (also spelt Occam), an English Franciscan friar and philosopher, c. 1287–1347.

training data. However, practice showed that even though the global optimum might not be found, the resulting GP based on suboptimal hyperparameters still explains the data well enough [52]. An alternative to evidence maximization is cross-validation, which is typically employed in SVM training. Cross-validation however is computationally demanding, requires MC simulation, and exploits only a fraction of the entire training data.

4.3 Large Data Sets

The computational complexity for calculating the posterior mean (4.6) and variance (4.7) is mainly dominated by the inversion of the matrix $\mathbf{K}_y$ and evaluating matrix-vector products. Inverting $\mathbf{K}_y$ or the more numerically robust alternative of computing its Cholesky decomposition is in $\mathcal{O}(n^3)$. Thus, for a large number of data points, this calculation becomes very demanding. Fortunately, the Cholesky decomposition of $\mathbf{K}_y$ can be determined off-line if the entire data set is given a priori. This, however is not the case for *streaming data*, where the data becomes available on-line during run-time. In this case, the matrix and thus its Cholesky decomposition has to be updated. In order to avoid a complete recalculation of the Cholesky decomposition, so-called *rank-1 Cholesky updates* can be performed for introducing new data points [142].

Given the inverse or Cholesky decomposition of $\mathbf{K}_y$, the complexity of the posterior mean is in $\mathcal{O}(n)$ and of the posterior variance is in $\mathcal{O}(n^2)$. For small data sets, these computations are affordable, but for large problems—tens of thousands data points—both storing $\mathbf{K}_y$ or its decomposition as well as solving the matrix-vector products in (4.6) and (4.7) is prohibitive. To overcome this problem, many approximate GP regression methods have been proposed, where some of the most popular are briefly reviewed in the following.

4.3.1 Active Set Approaches

In [141] a unifying framework for so-called *active set* approaches has been derived. Here, instead of processing the entire training data set, only a subset of the data points—the active set with $m \ll n$ data points—is used. The matrix $\mathbf{K}_y$ is replaced by a matrix that is based on the active set, which reduces the computational complexity to $\mathcal{O}(m^2 \cdot n)$ for matrix computations, $\mathcal{O}(m)$ for calculating the posterior mean, and $\mathcal{O}(m^2)$ for the posterior variance.

Several active set approaches have been proposed independently, but by means of the unifying framework in [141] it can be shown that all these approaches merely differ on modifications of the original GP prior (4.2). The *subset of regressors* approach [176, 205] is probably the simplest realization of an active set method, where the active set is either chosen randomly from the training set or by means of a greedy selection algorithm. The *sparse on-line GP* (SOGP) regression proposed in [48] in addition allows adding and removing data points to the active set at runtime, which is desirable when the training data set is not given entirely a priori. The *sparse pseudo-input GP* (SPGP, [177]) is generally regarded as one of the most effective active set approaches. In contrast to other methods, the active set has not to be a subset of the training data: elements of the active set can be located anywhere in the input domain. For this purpose, the active set is optimized by means of evidence maximization (4.12).

4.3.2 Local Approaches

To speed up GP regression, *local* or partitioning approaches split the training data into p data sets, where for each data set a separate (local) GP is learned. For calculating the posterior mean and variance, the individual estimates of the separate GPs are combined. In doing so, the complexity is $\mathcal{O}(p \cdot m^3)$ for matrix calculation, $\mathcal{O}(p \cdot m)$ for mean calculation, and $\mathcal{O}(p \cdot m^2)$ for calculating the variance, where $m = p/n$ is the number of data points per set. In contrast to active set methods, local approaches make use of the entire training data.

Two instances of a local approach are the *Bayesian committee machine* proposed in [199] as well as the *product of GP experts* proposed in [35]. In both methods partitioning is performed off-line. In contrast, the *local GP* [132] is an on-line algorithm, where incoming data is assigned to the nearest data set. To update the local kernel matrix, the above mentioned rank-1 updating is utilized.

4.3.3 Algebraic Tricks

Besides the above approximate regression approaches, there exist several approximations that tackle the algebraic operations necessary to compute the posterior mean and variances. This "algebraic tricks" can be used in combination with the above methods in order to further speed up GP regression. *Skilling's method* [175] for instance is an iterative procedure to solve matrix-vector operations of the kind $\mathbf{K}^{-1} \cdot \underline{x}$, which appear in both (4.6) and (4.7). If this method is terminated after k iterations, the complexity is $\mathcal{O}(k \cdot n^2)$ instead of $\mathcal{O}(n^3)$.

If the covariance function possesses as special structure, sparse versions of the Matrix $\mathbf{K}_y$ can be obtained allowing for efficient matrix-operations. This holds for covariance functions with compact support like the Matérn kernel. In case of a compact support, the complexity of the aforementioned rank-1 update reduces from $\mathcal{O}(n^2)$ down to $\mathcal{O}(n)$ [142]. In case of stationary covariance functions, the function values merely depend on the distance between pairs of inputs. By defining a distance threshold, only those pairs of training data points are considered that are within the threshold [203]. In addition, the distance dependency allows storing the training set in a *kd-tree*, which provides a rapid access to the data [73, 203].

4.3.4 Open Issues

Most of the proposed approximations for large data sets are not suitable for streaming data, which requires performing GP regression on-line at runtime. Among the on-line approaches, those utilizing rank-1 updating or partitioning merely alleviate the complexity problem. They still suffer from an increasing amount of data points, i.e., the computational and memory demand grows over time.

4.4 Nonlinear Filtering

Given a GP model of the latent function $g(.)$, it is so far possible to calculate the mean and variance for a deterministic input $\underline{x}$ according to (4.6) and (4.7), respectively. In order to perform Gaussian filtering, however, the input—more precisely the state—has to be a random vector. Accordingly, one has to solve the moment integral (2.8) for $\boldsymbol{g} \sim \mathcal{GP}(\mu, \kappa)$. As $\boldsymbol{g}$ is a random vector, it is in addition necessary to marginalize out the uncertainty over $g(.)$ in order to obtain the desired moments. Thus, for Gaussian filtering with GP models, the moment integrals (2.8) become

$$\begin{aligned}
\mu_y &= \iint g(\underline{x}) \cdot \mathcal{N}\left(\underline{x}; \underline{\mu}_x, \mathbf{C}_x\right) \cdot \mathcal{N}\left(g; \mu_g(\underline{x}), \sigma_g^2(\underline{x})\right) \mathrm{d}g\, \mathrm{d}\underline{x}\,, \\
\sigma_y^2 &= \iint \left(g(\underline{x}) - \mu_y\right)\left(g(\underline{x}) - \mu_y\right)^{\mathrm{T}} \cdot \\
&\qquad\qquad \mathcal{N}\left(\underline{x}; \underline{\mu}_x, \mathbf{C}_x\right) \cdot \mathcal{N}\left(g; \mu_g(\underline{x}), \sigma_g^2(\underline{x})\right) \mathrm{d}g\, \mathrm{d}\underline{x} + \sigma^2\,, \\
\sigma_{xy} &= \iint \left(\underline{x} - \underline{\mu}_x\right)\left(g(\underline{x}) - \mu_y\right)^{\mathrm{T}} \cdot \mathcal{N}\left(\underline{x}; \underline{\mu}_x, \mathbf{C}_x\right) \cdot \mathcal{N}\left(g; \mu_g(\underline{x}), \sigma_g^2(\underline{x})\right) \mathrm{d}g\, \mathrm{d}\underline{x}\,,
\end{aligned} \tag{4.13}$$

containing the additional integration over the latent function/GP $g(.)$. Here, the Gaussian density $\mathcal{N}(g; \mu_g(\underline{x}), \sigma_g^2(\underline{x}))$ is extracted from the GP, where the mean and variance are according to (4.6) and (4.7), respectively.

The number of approaches for Gaussian filtering with GP models proposed so far is limited. For arbitrary covariance functions, [69] proposed a first-order and second-order Taylor-series expansion of the posterior mean (4.6) and variance (4.7), respectively. Similarly, a first-order Taylor-series expansion of (4.6) was used in [104], which yields the so-called *GP extended Kalman filter* (GP-EKF). In [105], the unscented transform (see Section 2.2.5) is employed instead. The sample points are propagated through (4.6) and the desired moments (4.13) are then calculated by means of sample mean and variance, respectively. Accordingly, the Gaussian filter is named *GP unscented Kalman filter* (GP-UKF). In [68] it is shown, that for the SE covariance function, the mean μ_y and the variance σ_y^2 can be determined analytically. But in contrast to [104, 105], no full Gaussian filter that also involves a measurement update was provided. Furthermore, only a scalar output $\boldsymbol{y}$ is supported.

4.5 Contributions

In Section 4.5.1 and Section 4.5.2, a novel GP filtering and smoothing algorithm is proposed that is based on analytic moment matching, i.e., the moment integrals are solved exactly given a GP representation of the latent function $g(.)$. These results are based on Paper H and Paper I. An approximate on-line GP regression algorithm with on-line hyperparameter learning that was proposed in Paper J and Paper K is summarized in Section 4.5.3 and Section 4.5.4, respectively.

4.5.1 Gaussian Process Filtering

So far, only one-dimensional outputs $\boldsymbol{y}$ have been considered. For real-world applications however, the output is typically of higher dimension. Thus, in the following the focus is on a nonlinear transformation as in (2.6) with multivariate output $\underline{\boldsymbol{y}} \in \mathbb{R}^{n_y}$ and multivariate noise $\underline{\boldsymbol{w}} \sim \mathcal{N}(\underline{0}, \mathbf{C}_w)$ with $\mathbf{C}_w = \mathrm{diag}\big([\sigma_1^2 \ldots \sigma_{n_y}^2]\big)$. It is assumed that a separate GP $\mathcal{GP}(\mu_a, \kappa_a)$

is trained for each dimension $a = 1 \ldots n_y$. For training of the ath GP, the training data set $\mathcal{D}_a = \{\mathbf{X}_\mathcal{D}, \underline{y}_a\}$ consisting of n data points $(\underline{x}_i, y_{a,i})$, $i = 1 \ldots n$ is used.

Mean Vector $\underline{\mu}_y$

To obtain a Gaussian filter in this setup, the moment integrals (4.13) have to be solved. Due to the representation of the each dimension by means of a separate GP, the calculation of the mean vector $\underline{\mu}_y$ is also performed dimension-wise. The ath element $\mu_{y,a}$ of $\underline{\mu}_y$ is given by

$$\begin{aligned}\mu_{y,a} &= \int \mu_{g,a}(\underline{x}) \cdot \mathcal{N}\left(\underline{x}; \underline{\mu}_x, \mathbf{C}_x\right) \mathrm{d}\underline{x} \\ &= \sum_{i=1}^{n} \beta_{a,i} \cdot \int \kappa_a(\underline{x}, \underline{x}_i) \cdot \mathcal{N}\left(\underline{x}; \underline{\mu}_x, \mathbf{C}_x\right) \mathrm{d}\underline{x}\end{aligned} \tag{4.14}$$

with $\mu_{g,a}(.)$ being the ath element of the posterior mean (4.6) and $\beta_{a,i}$ being the ith element of the vector $\underline{\beta}_a \triangleq (\mathbf{K}_a + \sigma_a^2 \cdot \mathbf{I}_n)^{-1} \cdot \underline{y}_a$ with kernel matrix $\mathbf{K}_a$ consisting of the elements $(\mathbf{K}_a)_{e,f} = \kappa_a(\underline{x}_e, \underline{x}_f)$ for all $\underline{x}_e, \underline{x}_f \in \mathbf{X}_\mathcal{D}$. The second equality in (4.14) follows from writing the posterior mean function $\mu_{g,a}(.)$ as a finite sum over the covariance functions [144].

In general the integral in (4.14) cannot be solved in closed form due to the nonlinear covariance function, but if the covariance function belongs to one of the function types listed in Section 2.5.2, an analytical solution is possible. This holds for instance for the SE function (4.8) or the polynomial function (4.9). For the SE covariance function with signal variance α_a^2 and matrix of characteristic length-scales Λ_a, (4.14) can be simplified to

$$\mu_{y,a} = \underline{\beta}_a^{\mathrm{T}} \cdot \underline{q}_a \tag{4.15}$$

with $\underline{q}_a$ comprising the elements

$$q_{a,i} \triangleq \alpha_a^2 \cdot \left|\mathbf{C}_x \Lambda_a^{-1} + \mathbf{I}\right|^{-\frac{1}{2}} \cdot \exp\left(-\tfrac{1}{2}\left(\underline{x}_i - \underline{\mu}_x\right)^{\mathrm{T}} \left(\mathbf{C}_x + \Lambda_a\right)^{-1} \left(\underline{x}_i - \underline{\mu}_x\right)\right) , \tag{4.16}$$

for $i = 1 \ldots n$. A similar solution can be obtained for the polynomial covariance function (4.9).

Covariance Matrix $\mathbf{C}_y$

For calculating the covariance $\mathbf{C}_y$, again each element of the matrix is determined individually. At first, the off-diagonal elements $a, b = 1 \ldots n_y$ with $a \neq b$ of $\mathbf{C}_y$, i.e., the cross-covariances, are considered. These elements are given by

$$\begin{aligned} \sigma_{y,ab}^2 &= \int \mu_{g,a}(\underline{x}) \cdot \mu_{g,b}(\underline{x}) \cdot \mathcal{N}\left(\underline{x}; \underline{\mu}_x, \mathbf{C}_x\right) \mathrm{d}\underline{x} - \mu_{y,a} \cdot \mu_{y,b} \\ &= \sum_{i=1}^{n} \sum_{j=1}^{n} \beta_{a,i} \cdot \beta_{b,j} \cdot \\ &\qquad \int \kappa_a(\underline{x}, \underline{x}_i) \cdot \kappa_b(\underline{x}, \underline{x}_j) \cdot \mathcal{N}\left(\underline{x}; \underline{\mu}_x, \mathbf{C}_x\right) \mathrm{d}\underline{x} - \mu_{y,a} \cdot \mu_{y,b} , \end{aligned}$$

with $\mu_{y,a}$ and $\mu_{y,b}$ according to (4.15). For SE covariance functions, the cross-covariance can be further simplified to

$$\sigma_{y,ab}^2 = \underline{\beta}_a^{\mathrm{T}} \cdot \mathbf{Q} \cdot \underline{\beta}_b - \mu_{y,a} \cdot \mu_{y,b} \tag{4.17}$$

with elements of $\mathbf{Q} \in \mathbb{R}^{n \times n}$ according to

$$\begin{aligned} Q_{ij} &= \frac{\exp\left(n_{ij}^2\right)}{\sqrt{|\mathbf{R}|}} , \\ n_{ij}^2 &= \log\left(\alpha_a^2\right) + \log\left(\alpha_b^2\right) - \tfrac{1}{2}\left(\underline{\zeta}_i^{\mathrm{T}} \Lambda_a^{-1} \underline{\zeta}_i + \underline{\zeta}_j^{\mathrm{T}} \Lambda_b^{-1} \underline{\zeta}_j - \underline{z}_{ij}^{\mathrm{T}} \mathbf{R}^{-1} \mathbf{C}_x \underline{z}_{ij}\right) , \end{aligned} \tag{4.18}$$

where $\mathbf{R} \triangleq \mathbf{C}_x\left(\Lambda_a^{-1} + \Lambda_b^{-1}\right) + \mathbf{I}$, $\underline{\zeta}_i \triangleq \underline{x}_i - \underline{\mu}_x$, and $\underline{z}_{ij} \triangleq \Lambda_a^{-1} \underline{\zeta}_i + \Lambda_b^{-1} \underline{\zeta}_j$.

The diagonal elements ($a = b$) of $\mathbf{C}_y$ comprise in addition to (4.17) a term reflecting the noise variance σ_a^2 and the uncertainty of the GP

$$\alpha_a^2 - \mathrm{Tr}\left(\left(\mathbf{K}_a + \sigma_a^2 \mathbf{I}\right)^{-1} \mathbf{Q}\right) + \sigma_a^2 \,. \tag{4.19}$$

Hence, the desired elements of covariance matrix $\mathbf{C}_y$ are given by

$$\sigma_{y,ab}^2 = \begin{cases} \text{Eq. (4.17)} + \text{Eq. (4.19)} & \text{if } a = b \\ \text{Eq. (4.17)} & \text{otherwise} \end{cases} \,. \tag{4.20}$$

Cross-Covariance Matrix $\mathbf{C}_{xy}$

It remains to compute the cross-covariance $\mathbf{C}_{xy}$ to fully determine a Gaussian filter for GP models. Integrating out $\underline{g}$, the cross-covariance can be simplified to

$$\mathbf{C}_{xy} = \int \underline{x} \cdot \left(\underline{\mu}_g(\underline{x})\right)^{\mathrm{T}} \cdot \mathcal{N}\left(\underline{x}; \underline{\mu}_x, \mathbf{C}_x\right) \mathrm{d}\underline{x} - \underline{\mu}_x \cdot \underline{\mu}_y^{\mathrm{T}} \,.$$

By writing $\underline{\mu}_g(\underline{x})$ as a finite sum over covariance functions, the ath column of $\mathbf{C}_{xy}$, $a = 1 \ldots n_y$ can be written as

$$\sum_{i=1}^{n} \beta_{a,i} \cdot \int \underline{x} \cdot \kappa_a(\underline{x}, \underline{x}_i) \cdot \mathcal{N}\left(\underline{x}; \underline{\mu}_x, \mathbf{C}_x\right) \mathrm{d}\underline{x} - \underline{\mu}_x \cdot \mu_{y,a} \,. \tag{4.21}$$

With a SE covariance function, this term can be solved analytically to

$$\sum_{i=1}^{n} \beta_{a,i} \cdot q_{a,i} \cdot \mathbf{C}_x \cdot \left(\mathbf{C}_x + \Lambda_a\right)^{-1} \cdot \left(\underline{x}_i - \underline{\mu}_x\right) , \tag{4.22}$$

where the analytical solution (4.15) for $\underline{\mu}_{y,a}$ has been substituted and $q_{a,i}$ coincides with (4.16).

The Gaussian filter for GP models based on the expressions (4.15), (4.20), and (4.22) for calculating the moments in (4.13) is named *GP assumed density filter* (GP-ADF) in the following.

Example 20: One Step Filtering

The nonlinear dynamic system

$$\boldsymbol{x}_k = \frac{\boldsymbol{x}_{k-1}}{2} + \frac{25\boldsymbol{x}_{k-1}}{1+\boldsymbol{x}_{k-1}^2} + \boldsymbol{w}_k\,, \qquad \text{with } \boldsymbol{w}_k \sim \mathcal{N}\left(0, \sigma_w^2 = 0.2^2\right), \tag{4.23}$$

$$\boldsymbol{z}_k = 5 \cdot \sin(\boldsymbol{x}_k) + \boldsymbol{v}_k\,, \qquad \text{with } \boldsymbol{v}_k \sim \mathcal{N}\left(0, \sigma_v^2 = 0.2^2\right), \tag{4.24}$$

is considered, which is a modified version of the model used in [60, 102]. The standard deviation of the initial state is set to be $\sigma_0^x = 0.5$, i.e., the initial uncertainty is fairly high. The system and measurement noises are relatively small considering the amplitudes of the system function and the measurement function. For the numerical analysis, the mean values $\mu_{0,i}^x$, $i = 1 \ldots 100$, are placed equidistantly on the interval $[-3,3]$. Then, a single (initial) state $x_{0,i}$ is sampled from $\mathcal{N}\left(\mu_{0,i}^x, \left(\sigma_0^x\right)^2\right)$, $i = 1 \ldots 100$.

For the dynamic system in (4.23)–(4.24), the performance of a single prediction and measurement update of the EKF, the UKF, the CKF, the GP-UKF, and the GP-ADF is compared against the ground truth, which is approximated by means of a near-optimal sampling-based Gaussian filter (denoted as *Gibbs-filter*, [55]) as well as a PF with 200 particles. Compared to the evaluation of longer trajectories, evaluating a single filtering step makes it easier to analyze the estimates of individual filtering algorithms.

Table 4.1 summarizes the performances (rmse, mae, nll) of all filters for estimating the latent state $\boldsymbol{x}$. The results in the table are based on averages over 1,000 test runs and 100 randomly sampled initial states per test run. The table also reports the 95% standard error of the expected performances.

Table 4.1: Average performances (rmse, mae, nll) with standard errors (95% confidence interval) and p-values testing the hypothesis that the other filters are better than the GP-ADF using a one-sided t-test. The green color highlights the near-optimal results of the Gibbs-filter and the PF.

	rmse		mae		nll	
	average	p-value	average	p-value	average	p-value
EKF	3.62 ± 0.212	4.1×10^{-2}	2.36 ± 0.176	0.38	$3.05 \times 10^{3} \pm 3.02 \times 10^{2}$	$< 10^{-4}$
UKF	10.5 ± 1.08	$< 10^{-4}$	8.58 ± 0.915	$< 10^{-4}$	25.6 ± 3.39	$< 10^{-4}$
CKF	9.24 ± 1.13	2.8×10^{-4}	7.31 ± 0.941	4.2×10^{-4}	$2.22 \times 10^{2} \pm 17.5$	$< 10^{-4}$
GP-UKF	5.36 ± 0.461	7.9×10^{-4}	3.84 ± 0.352	3.3×10^{-3}	6.02 ± 0.497	$< 10^{-4}$
GP-ADF	$\mathbf{2.85 \pm 0.174}$	—	$\mathbf{2.17 \pm 0.151}$	—	$\mathbf{1.97 \pm 6.55 \times 10^{-2}}$	—
Gibbs	2.82 ± 0.171	0.54	2.12 ± 0.148	0.56	$1.96 \pm 6.62 \times 10^{-2}$	0.55
PF	$1.57 \pm 7.66 \times 10^{-2}$	1.0	$0.36 \pm 2.28 \times 10^{-2}$	1.0	$1.03 \pm 7.30 \times 10^{-2}$	1.0

Table 4.1 indicates that the GP-ADF is the most robust filter and statistically significantly outperforms all filters but the Gibbs-filter and the PF. Amongst all other filters the GP-ADF is the closest Gaussian filter to the computationally expensive Gibbs-filter [55]. Note that the PF is not a Gaussian filter and is able to express multi-modality in densities. Therefore, its performance is typically better than the one of Gaussian filters. The difference between the PF and a near-optimal Gaussian filter, the Gibbs-filter, is expressed in Table 4.1. The performance difference essentially depicts how much is lost by using a Gaussian filter instead of a particle filter.

The poor performance of the EKF is due to linearization errors. The filters based on small sample approximations of densities (UKF, GP-UKF, CKF) suffer from the degeneracy of these approximations, which is illustrated in Figure 4.3. Note that the CKF uses a smaller set of sample points than the UKF (recall Section 2.2.5), which makes the CKF statistically even less robust than the UKF.

4.5.2 Gaussian Process Smoothing

For an RTSS given a GP representation of the system and measurement function, most of the ingredients are already derived by means of the above GP-ADF. The only missing component is the cross-covariance matrix $\mathbf{C}_{k|k+1}$ in (2.33). For its calculation, one can follow the above derivation of the cross-covariance matrix $\mathbf{C}_{xy}$. According to (4.21), the ath column of $\mathbf{C}_{k|k+1}$ can be written as

$$\sum_{i=1}^{n} \beta_{a,i} \cdot \int \underline{x} \cdot \kappa_a(\underline{x}, \underline{x}_i) \cdot \mathcal{N}\left(\underline{x}; \underline{\mu}_k^e, \mathbf{C}_k^e\right) \mathrm{d}\underline{x} - \underline{\mu}_k^e \cdot \mu_{k,a}^p , \tag{4.25}$$

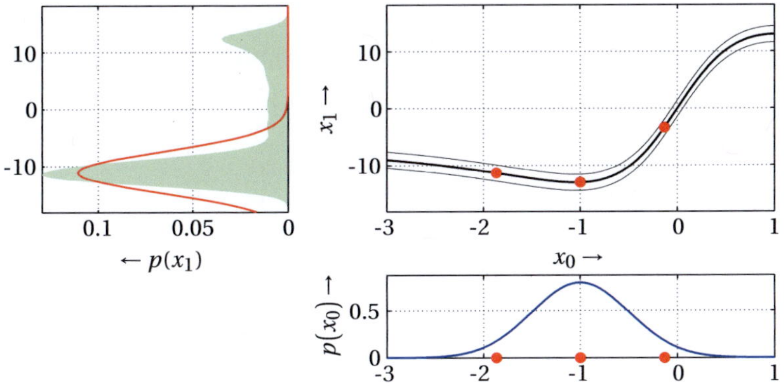

(a) UKF time update $p(x_1|\emptyset)$, which misses out substantial probability mass of the true predictive distribution.

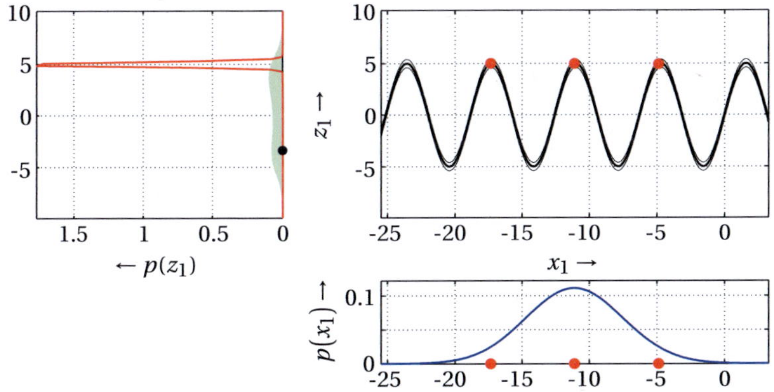

(b) UKF determines $p(z_1|\emptyset)$, which is too sensitive and cannot explain the actual measurement z_1 (black dot, left sub-figure).

Figure 4.3: Degeneracy of the unscented transformation (UT) underlying the UKF. Input distributions to the UT are the Gaussians in the sub-figures at the bottom in each panel. The functions the UT is applied to are shown in the top right sub-figures, i.e, the transition mapping (4.23) in (a) and the measurement mapping (4.24) in (b). Sigma points are marked by red dots. The predictive distributions are shown in the left sub-figures of each panel. The true predictive distributions are the shaded areas; the UT predictive distributions are the solid Gaussians. The predictive distribution of the time update in (a) equals the input distribution at the bottom of (b).

with $\mathcal{N}\left(\underline{x}; \underline{\mu}_k^e, \mathbf{C}_k^e\right)$ being the posterior state density and $\underline{\mu}_{k,a}^p$ being the ath element of the predicted mean $\underline{\mu}_k^p$. For an SE covariance function, a closed-form expression of (4.25) can be obtain given by

$$\sum_{i=1}^{n} \beta_{a,i} \cdot q_{a,i} \cdot \mathbf{C}_k^e \cdot \left(\mathbf{C}_k^e + \Lambda_a\right)^{-1} \cdot \left(\underline{x}_i - \underline{\mu}_k^e\right) , \tag{4.26}$$

with $q_{a,i}$ as in (4.16), but with $\mathbf{C}_x$ and $\underline{\mu}_x$ being substituted with $\mathbf{C}_k^e$ and $\underline{\mu}_k^e$, respectively.

Altogether, the *GP Rauch-Tung-Striebel smoother* (GP-RTSS) utilizes the GP-ADF of the previous section for the forward sweep (prediction and measurement update). For the actual smoothing, (4.26) is used for calculating the cross-covariance $\mathbf{C}_{k|k+1}$, which is required for determining the smoothed Gaussian state according to (2.34).

The computational complexity of prediction, measurement update, and smoothing (after training the GPs) is in $\mathcal{O}\left(K \cdot n^2 \cdot \left(n_x^3 + n_z^3\right)\right)$ due to matrix inversions, matrix multiplications, and the computation of the $\mathbf{Q}$-matrix (4.18). For comparison, Kalman filter and RTSS scale with $\mathcal{O}\left(K \cdot \left(n_x^3 + n_z^3\right)\right)$.

Example 21: Pendulum Tracking

The pendulum tracking example taken from [53] is considered for comparing the performances of four filters and smoothers: the EKF/EKS, the UKF/URTSS, the GP-UKF/GP-URTSS, the CKF/CKS, the Gibbs-filter/smoother, and the GP-ADF/GP-RTSS. The pendulum has mass $m = 1\,\mathrm{kg}$ and length $l = 1\,\mathrm{m}$. The state $\underline{\boldsymbol{x}} = \left[\dot{\boldsymbol{\varphi}}\ \boldsymbol{\varphi}\right]^{\mathrm{T}}$ of the pendulum is given by the angle $\boldsymbol{\varphi}$ (measured anti-clockwise from hanging down) and the angular velocity $\dot{\boldsymbol{\varphi}}$. The pendulum can exert a constrained torque $u \in [-5,5]\,\mathrm{Nm}$. A frictionless system is assumed such that the system function $\underline{a}(.)$ is

$$\underline{a}\left(\underline{x}_k, u_k\right) = \int_k^{k+\Delta_k} \begin{bmatrix} \frac{u(\tau) - 0.5\, m l g \sin\left(\varphi(\tau)\right)}{0.25\, m l^2 + I} \\ \dot{\varphi}(\tau) \end{bmatrix} \mathrm{d}\tau , \tag{4.27}$$

Table 4.2: Averaged filtering and smoothing performances with 95% confidence intervals.

Filters	nll	Smoothers	nll
EKF	$1.6 \times 10^2 \pm 29.1$	EKS [121]	$\mathbf{3.3 \times 10^2 \pm 60.5}$
UKF	6.0 ± 3.02	URTSS [159]	$\mathbf{17.2 \pm 10.0}$
CKF	28.5 ± 9.83	CKS [55]	$\mathbf{72.0 \pm 25.1}$
GP-UKF$_{250}$	4.4 ± 1.32	GP-URTSS$_{250}$ [54]	$\mathbf{10.3 \pm 3.85}$
GP-ADF$_{250}$	$\mathbf{1.44 \pm 0.117}$	GP-RTSS$_{250}$	$\mathbf{1.04 \pm 0.204}$
GP-ADF$_{20}$	6.63 ± 0.149	GP-RTSS$_{20}$	6.57 ± 0.148

where I is the moment of inertia and g the acceleration of gravity. Then, the successor state

$$\underline{\boldsymbol{x}}_{k+1} = \underline{\boldsymbol{x}}_{k+\Delta_k} = \underline{a}(\underline{\boldsymbol{x}}_k, u_k) + \underline{\boldsymbol{w}}_k\,, \quad \text{with } \underline{\boldsymbol{w}}_k \sim \mathcal{N}\big(\underline{0}, \text{diag}\big(0.5^2, 0.1^2\big)\big)$$

is computed using an ODE solver for (4.27) with a zero-order hold control signal $u(\tau)$. The torque is sampled randomly according to $u \sim \mathcal{U}[-5,5]$ Nm and implemented using a zero-order-hold controller. Every time increment $\Delta_k = 0.2$ s, the state is observed according to

$$\boldsymbol{z}_k = \arctan\left(\frac{-1-l\cdot\sin(\boldsymbol{\varphi}_k)}{0.5-l\cdot\cos(\boldsymbol{\varphi}_k)}\right) + \boldsymbol{v}_k\,, \quad \text{with } \boldsymbol{v}_k \sim \mathcal{N}\big(0, 0.05^2\big)\,. \tag{4.28}$$

Trajectories of length $K = 6\,\text{s} = 30$ time steps are started from a state sampled from $\mathcal{N}\big(\underline{\mu}_0^x, \mathbf{C}_0^x\big)$ with mean $\underline{\mu}_0^x = [0\;0]^{\mathrm{T}}$ and covariance $\mathbf{C}_0^x = \text{diag}\big(0.01^2, (\pi/16)^2\big)$. For each trajectory, GP models $\mathcal{GP}_a$ (system function) and $\mathcal{GP}_h$ (measurement function) were learned based on randomly generated data using either 250 or 20 data points.

Table 4.2 reports the values of the nllmeasure for the EKF/EKS, the UKF/URTSS, the GP-UKF/GP-URTSS, the GP-ADF/GP-RTSS, and the CKF/CKS, averaged over 1,000 MC runs. The GP-RTSS is the only method that consistently reduces the nll value compared to the corresponding filtering algorithm. Increased nll values (red color in Table 4.2) occur when the state density cannot explain the state/measurement. A detailed example of this can be found in Paper I. Even with only 20 training points, the GP-ADF/GP-RTSS outperforms the EKF/EKS, UKF/URTSS, CKF/CKS.

4.5.3 Recursive Gaussian Process Regression

As mentioned in Section 4.3, the complexity of GP regression scales cubically with the number of training data points. This complexity can be reduced by means of sparse approximations. Most of these approximations however only work in an off-line mode, i.e., all training data has to be known a priori and is processed in a batch. This is not suitable for streaming data. The *recursive Gaussian Process* (RGP) regression approach introduced next allows for both a sparse representation *and* on-line processing. For this purpose, the latent function is represented by means of a finite set of so-called *basis vectors.*

Let $\mathbf{X} \triangleq [\underline{x}_1, \underline{x}_2, \ldots, \underline{x}_m]$ be the matrix of locations of the basis vectors, where the number of basis vectors m is significantly lower than the size n of $\mathcal{D}$, i.e., $m \ll n$. Furthermore, $\underline{\mathrm{g}} \triangleq g(\mathbf{X})$ are the (unkown) values of the latent function at the locations $\mathbf{X}$. The basis vectors can be considered an active set allowing a sparse GP representation. In contrast to most other active set approaches, the basis vectors are updated *on-line* with new observations $\underline{y}_k$ at inputs $\mathbf{X}_k \triangleq \left[\underline{x}_{k,1}, \underline{x}_{k,2}, \ldots, \underline{x}_{k,n_k}\right]$ and time step $k = 0, 1, \ldots$, which facilitates to process streaming data. Hence, $\underline{y}_k$ and $\mathbf{X}_k$ can be considered a subset of $\underline{y}_{\mathcal{D}}$ and $\mathbf{X}_{\mathcal{D}}$, respectively, but where $\underline{y}_{\mathcal{D}}$ and $\mathbf{X}_{\mathcal{D}}$ are not known completely a priori. Also off-line processing is possible by presenting $\underline{y}_{\mathcal{D}}$ and $\mathbf{X}_{\mathcal{D}}$ in batches to the algorithm.

For all steps $k = 0, 1, \ldots$ it is assumed that the basis vectors are fixed in number and location. Since $g(\underline{x})$ is assumed to be a GP, the initial distri-

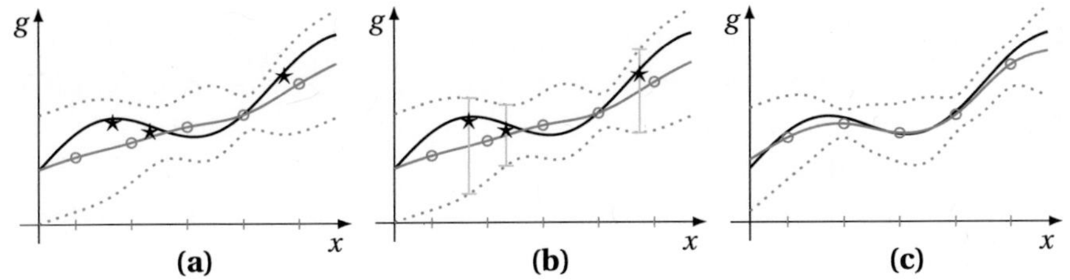

Figure 4.4: (a) The black line indicates the latent function g, while the gray solid and dotted lines represent the mean and variance of recursive GP. The circles indicate the location of the basis vectors $\mathbf{X}$ (x-axis) and their mean values $\underline{\mu}^g$ (y-axis). The stars indicate new observations. (b) Inferring the mean and covariance of g at the locations of the new observations from the current recursive GP estimate. (c) Updating the GP with the new observations gives an improved estimate of the latent function.

bution $f_0(\underline{g}) = \mathcal{N}\left(\underline{g}; \underline{\mu}_0^g, \mathbf{C}_0^g\right)$ of $\underline{\mathbf{g}}$ for $k = 0$ is Gaussian with mean vector $\underline{\mu}_0^g \triangleq \mu(\mathbf{X})$ and covariance matrix $\mathbf{C}_0^g \triangleq \kappa(\mathbf{X},\mathbf{X})$ according to (4.2).

The goal is now to calculate the posterior distribution $f\left(\underline{g}\middle|\underline{y}_{1:k}\right)$ recursively by updating the prior distribution of $\underline{g}$ from the previous step $k-1$

$$f_{k-1} \triangleq f_{k-1}\left(\underline{g}\middle|\underline{y}_{1:k-1}\right) = \mathcal{N}\left(\underline{g}; \underline{\mu}_{k-1}^g, \mathbf{C}_{k-1}^g\right) \tag{4.29}$$

with the new observations $\underline{y}_k$. For this purpose, the desired posterior distribution is expanded according to

$$f_k = \int \underbrace{c_k \cdot f\left(\underline{y}_k\middle|\underline{g}, \underline{g}_k\right) \cdot f\left(\underline{g}, \underline{g}_k\middle|\underline{y}_{1:k-1}\right)}_{= f\left(\underline{g}, \underline{g}_k\middle|\underline{y}_{1:k}\right)} \mathrm{d}\underline{g}_k \tag{4.30}$$

by applying Bayes' law and by integrating out $\underline{g}_k \triangleq g(\mathbf{X}_k)$ from the joint posterior $f\left(\underline{g}, \underline{g}_k\middle|\underline{y}_{1:k}\right)$. Here, c_k is a normalization constant. Based on (4.30), calculating the posterior distribution can be performed in two steps: *Inference*, i.e., calculating the joint prior $f\left(\underline{g}, \underline{g}_k \middle| \underline{y}_{1:k-1}\right)$ given the prior f_{k-1} in (4.29). *Update*, i.e., updating the joint prior with the observations $\underline{y}_k$ and integrating out $\underline{g}_k$. The interaction between both steps is depicted in Figure 4.4.

Inference

In order to determine the joint prior $f\left(\underline{g}, \underline{g}_k \middle| \underline{y}_{1:k-1}\right)$, it is important to emphasize that the joint distribution $f\left(\underline{g}, \underline{g}_k\right)$ is Gaussian with mean and covariance

$$\underline{\mu} = \begin{bmatrix} \mu(\mathbf{X}) \\ \mu(\mathbf{X}_k) \end{bmatrix} \quad \text{and} \quad \mathbf{C} = \begin{bmatrix} \kappa(\mathbf{X},\mathbf{X}) & \kappa(\mathbf{X},\mathbf{X}_k) \\ \kappa(\mathbf{X}_k,\mathbf{X}) & \kappa(\mathbf{X}_k,\mathbf{X}_k) \end{bmatrix} , \tag{4.31}$$

respectively. This follows from the fact that $g(.)$ is a GP and any finite representation of this GP yields a Gaussian distribution. Thus, the joint prior can be written as

$$f\left(\underline{g}, \underline{g}_k \middle| \underline{y}_{1:k-1}\right) \approx f\left(\underline{g}_k \middle| \underline{g}\right) \cdot f_{k-1} = \mathcal{N}\left(\underline{g}_k; \underline{\mu}_k^p, \mathbf{B}\right) \cdot \mathcal{N}\left(\underline{g}; \underline{\mu}_{k-1}^g, \mathbf{C}_{k-1}^g\right) , \tag{4.32}$$

with

$$\underline{\mu}_k^p = \mu(\mathbf{X}_k) + \mathbf{J}_k \cdot \left(\underline{\mu}_{k-1}^g - \mu(\mathbf{X})\right) , \tag{4.33}$$

$$\mathbf{B} = \kappa(\mathbf{X}_k,\mathbf{X}_k) - \mathbf{J}_k \cdot \kappa(\mathbf{X},\mathbf{X}_k) , \tag{4.34}$$

$$\mathbf{J}_k = \kappa(\mathbf{X}_k,\mathbf{X}) \cdot \kappa(\mathbf{X},\mathbf{X})^{-1} . \tag{4.35}$$

The first equality in (4.32) follows from assuming that $\underline{\mathbf{g}}_k$ is conditionally independent of the past observations $\underline{y}_{1:k-1}$ given $\underline{\mathbf{g}}$.[7] Hence, the conditional distribution $f\left(\underline{g}_k \middle| \underline{g}\right)$ is Gaussian and results from the joint

7 This is true if all inputs $\mathbf{X}_{1:k-1}$ of the past observations are a subset of the basis vectors $\mathbf{X}$, otherwise it is an approximation.

distribution $f\left(\underline{g},\underline{g}_k\right)$ in (4.31) by conditioning on $\underline{\mathbf{g}}$ (see (2.9)), which results in the second equality.

After some algebraic transformations, where some basic properties of Gaussian distributions and the Woodbury formula are utilized, the product in (4.32) yields the joint Gaussian $f\left(\underline{g},\underline{g}_k\middle|\underline{y}_{1:k-1}\right) = \mathcal{N}\left(\underline{q};\mathbf{Q}\right)$ of $\underline{\mathbf{g}}$ and $\underline{\mathbf{g}}_k$ with mean and covariance

$$\underline{q} \triangleq \begin{bmatrix} \underline{\mu}_{k-1}^{g} \\ \underline{\mu}_k^{p} \end{bmatrix} \quad \text{and} \quad \mathbf{Q} \triangleq \begin{bmatrix} \mathbf{C}_{k-1}^{g} & \mathbf{C}_{k-1}^{g}\mathbf{J}_k^{\mathrm{T}} \\ \mathbf{J}_k\mathbf{C}_{k-1}^{g} & \mathbf{C}_k^{p} \end{bmatrix}, \tag{4.36}$$

respectively, and with covariance $\mathbf{C}_k^p \triangleq \mathbf{B} + \mathbf{J}_k\mathbf{C}_{k-1}^{g}\mathbf{J}_k^{\mathrm{T}}$. For a detailed derivation see Paper J. A close inspection of the second row in (4.36) shows that it has the same structure as an RTSS (see Section 2.3) and it coincides with the augmented Kalman Smoother proposed in [149], but there no update step for basis vectors as introduced next is derived.

Update

Given the result of the previous section that the joint prior in (4.32) is a Gaussian $\mathcal{N}(\underline{q},\mathbf{Q})$, the next step is to perform the update and marginalization in (4.30). For this purpose, (4.30) is rearranged to

$$f_k = \int \underbrace{c_k \cdot f\left(\underline{y}_k\middle|\underline{g}_k\right) \cdot \overbrace{f\left(\underline{g}_k\middle|\underline{y}_{1:k-1}\right) \cdot f\left(\underline{g}\middle|\underline{g}_k,\underline{y}_{1:k-1}\right)}^{=f\left(\underline{g},\underline{g}_k\middle|\underline{y}_{1:k-1}\right)}}_{=f\left(\underline{g}_k\middle|\underline{y}_{1:k}\right)\ \text{(Kalman filter)}} \mathrm{d}\underline{g}_k \tag{4.37}$$

under consideration that $\underline{\mathbf{g}}$ is not observed and thus, $f\left(\underline{y}_k\middle|\underline{g}_k\right)$ is independent of $\underline{\mathbf{g}}$. Since $f\left(\underline{y}_k\middle|\underline{g}_k\right) = \mathcal{N}\left(\underline{y}_k;\underline{g}_k,\sigma^2\mathbf{I}\right)$ according to (4.5) and $f\left(\underline{g}_k\middle|\underline{y}_{1:k-1}\right) = \mathcal{N}\left(\underline{g}_k;\underline{\mu}_k^p,\mathbf{C}_k^p\right)$ are both Gaussian, $\underline{\mathbf{g}}_k$ can be updated easily via a *Kalman filter* update step. Updating $\underline{\mathbf{g}}$ and integrating out $\underline{g}_k$

Algorithm 4 Recursive Gaussian Process (RGP)

▷ *Inference*
1: Calculate gain matrix $\mathbf{J}_k$ according to (4.35)
2: Calculate mean $\underline{\mu}_k^p$ via (4.33) and covariance matrix $\mathbf{C}_k^p$ via (4.36)
▷ *Update*
3: Calculate gain matrix $\mathbf{G}_k$ according to (4.40)
4: Calculate mean $\underline{\mu}_k^g$ via (4.38) and covariance matrix $\mathbf{C}_k^g$ via (4.39)

is then performed simultaneously, which yields the desired posterior $f_k = \mathcal{N}\left(\underline{g}; \underline{\mu}_k^g, \mathbf{C}_k^g\right)$ with

$$\underline{\mu}_k^g = \underline{\mu}_{k-1}^g + \mathbf{G}_k \cdot \left(\underline{y}_k - \underline{\mu}_k^p\right) , \tag{4.38}$$

$$\mathbf{C}_k^g = \mathbf{C}_{k-1}^g - \mathbf{G}_k \mathbf{J}_k \mathbf{C}_{k-1}^g , \tag{4.39}$$

$$\mathbf{G}_k = \mathbf{C}_{k-1}^g \mathbf{J}_k^{\mathrm{T}} \cdot \left(\mathbf{C}_k^p + \sigma^2 \mathbf{I}\right)^{-1} . \tag{4.40}$$

Putting all together, at steps $k = 1, 2, \ldots$ the proposed RGP recursively processes observations $\underline{y}_k$ at the inputs $\mathbf{X}_k$ as listed in Algorithm 4. This recursion commences from the initial mean $\underline{\mu}_0^g = \mu(\mathbf{X})$ and covariance $\mathbf{C}_0^g = \kappa(\mathbf{X}, \mathbf{X})$.

Discussion

So far, it was assumed that the set of basis vectors is fixed. The inference step, however, can also be utilized for introducing new basis vectors $\mathbf{X}'$. This might be of interest in locations where the current estimate of the latent function is inaccurate. By replacing $\mathbf{X}_k$ with $[\mathbf{X}, \mathbf{X}']$, the inference step provides the initial mean and covariance as well as the cross-covariance between the new basis vectors and the old ones.

The computations of the inference step scale with $\mathcal{O}(m^2 \cdot n_k)$ due to calculating $\mathbf{J}_k$ in (4.35), where n_k is the number of observations at step k.

Here, the inversion of the kernel matrix $\kappa(\mathbf{X},\mathbf{X})$ is computationally unproblematic, as it has to be calculated only once at step $k = 0$. Once the gain matrix $\mathbf{J}_k$ is calculated, predictions for a single test input are in $\mathcal{O}(m)$ (mean) and $\mathcal{O}(m^2)$ (covariance). Assuming that all observations are processed at once, predictions of the RGP are as complex as predictions of active set GP approaches. In contrast to most sparse GP approaches, the proposed method can process new observations on-line.

The update step scales with $\mathcal{O}(n_k \cdot m^2)$, where the complexity results from matrix multiplications for which more efficient algorithms exist, e.g., Strassen's algorithm [187]. The inversion in (4.40) again is not critical as the affected matrix is of size $n_k \times n_k$, where typically $n_k \ll m$.

4.5.4 On-line Hyperparameter Learning

In the following, the previous assumption of a-priori known hyperparameters is relaxed. Instead, the goal is now to learn the hyperparameters $\underline{\theta} \in \mathbb{R}^{n_\theta}$ simultaneously with estimating the values of the latent function $g(.)$ at the basis vectors. This is achieved by formulating the learning part as a recursive parameter estimation problem, which can be performed together with the function value estimation. Similar to Section 4.5.3, this boils down to calculating a joint posterior $f_k \triangleq f\left(\underline{\xi}_k \middle| \underline{y}_{1:k}\right) = \mathcal{N}\left(\underline{\xi}_k; \underline{\mu}_k^\xi, \mathbf{C}_k^\xi\right)$, where $\underline{\boldsymbol{\xi}}_k^{\mathrm{T}} \triangleq \left[\underline{\mathbf{g}}^{\mathrm{T}}\ \underline{\boldsymbol{\theta}}_k^{\mathrm{T}}\right]$ is the joint hidden state with mean and covariance

$$\underline{\mu}_k^\xi \triangleq \begin{bmatrix} \underline{\mu}_k^g \\ \underline{\mu}_k^\eta \end{bmatrix}, \quad \mathbf{C}_k^\xi \triangleq \begin{bmatrix} \mathbf{C}_k^g & \mathbf{C}_k^{g\eta} \\ \mathbf{C}_k^{\eta g} & \mathbf{C}_k^\eta \end{bmatrix}.$$

Starting point for this calculation is a joint prior $f\left(\underline{\xi}_{k-1} \middle| \underline{y}_{1:k-1}\right)$ at step $k-1$, which is updated with the new observations $\underline{y}_k$. This requires the following two operations: *Inference*, i.e., calculating a joint density $f\left(\underline{\xi}_{k-1}, \underline{g}_k \middle| \underline{y}_{1:k-1}\right)$ by exploiting the results of Section 4.5.3, and *Update*, i.e., incorporation of the new observations $\underline{y}_k$ and marginalization to obtain f_k.

Inference

To incorporate the new inputs $\mathbf{X}_k$, it is necessary to infer the latent function $g(.)$ at $\mathbf{X}_k$. For this purpose, the intermediate result (4.36) is exploited. The part of the mean $\underline{q}$ and the covariance $\mathbf{Q}$ regarding $\underline{g}_k$ can alternatively be calculated by employing a Kalman predictor on the linear state-space model

$$\underline{\mathbf{g}}_k = \mathbf{J}_k \cdot \underline{\mathbf{g}} + \underline{\boldsymbol{w}}_k \quad , \quad \underline{\boldsymbol{w}}_k \sim \mathcal{N}(\underline{b}, \mathbf{B}) \,, \tag{4.41}$$

where $\underline{b} \triangleq m(\mathbf{X}_k) - \mathbf{J}_k \cdot m(\mathbf{X})$ and $\mathbf{B}$ is according to (4.34). In order to also correlate $\underline{\mathbf{g}}_k$ with the hyperparameters, the model in (4.41) is extended to a state-space model given by

$$\begin{bmatrix} \underline{\boldsymbol{\xi}}_{k-1} \\ \underline{\mathbf{g}}_k \end{bmatrix} = \begin{bmatrix} \mathbf{I} & \mathbf{0} \\ \mathbf{0} & \mathbf{I} \\ \mathbf{J}_k\left(\underline{\boldsymbol{\theta}}_{k-1}\right) & \mathbf{0} \end{bmatrix} \cdot \underbrace{\begin{bmatrix} \underline{\mathbf{g}} \\ \underline{\boldsymbol{\theta}}_{k-1} \end{bmatrix}}_{\underline{\xi}_{k-1}} + \underline{\boldsymbol{w}}_k \,, \tag{4.42}$$

where the noise $\underline{\boldsymbol{w}}_k \sim \mathcal{N}\left(\underline{\mu}_k^w, \mathbf{C}_k^w\right)$ is Gaussian with mean and covariance

$$\underline{\mu}_k^w \triangleq \begin{bmatrix} \underline{0} \\ \underline{0} \\ \underline{b}\left(\underline{\boldsymbol{\theta}}_{k-1}\right) \end{bmatrix} , \mathbf{C}_k^w \triangleq \begin{bmatrix} \mathbf{0} & \mathbf{0} & \mathbf{0} \\ \mathbf{0} & \mathbf{0} & \mathbf{0} \\ \mathbf{0} & \mathbf{0} & \mathbf{B}\left(\underline{\boldsymbol{\theta}}_{k-1}\right) \end{bmatrix} , \tag{4.43}$$

respectively. Here, the dependence on the hyperparameters has been made explicit. The first two rows in (4.42) and (4.43) are merely an identity mapping of the given joint state $\underline{\boldsymbol{\xi}}_{k-1}$, while the last row corresponds to (4.41).

Based on model (4.42), performing a prediction would yield the desired joint distribution $f\left(\underline{\xi}_{k-1}, \underline{g}_k \middle| \underline{y}_{1:k-1}\right)$. As the model is nonlinear with respect to the hyperparameters $\underline{\boldsymbol{\theta}}_{k-1}$, the prediction cannot be performed exactly in closed form. Fortunately, the model is conditionally linear. Thus, the prediction can be approximated efficiently and accurately by means of the decomposition technique proposed in Section 2.5.2, where the nonlinear part—the hyperparameters—are sampled by means of an LRKF while the prediction of $\underline{\boldsymbol{g}}$ can be performed exactly via the Kalman predictor.

Update

In order to incorporate the new observations $\underline{y}_k$ in a very computationally efficient manner, the update is performed by means of the observed-unobserved decomposition proposed in Section 2.5.1. The observed state $\underline{\boldsymbol{\xi}}_k^o$ comprises the noise standard deviation $\boldsymbol{\sigma}$ as well as $\underline{\boldsymbol{g}}_k$, while the unobserved state $\underline{\boldsymbol{\xi}}_k^u$ comprises $\underline{\boldsymbol{g}}$ and $\underline{\boldsymbol{\theta}}_k^-$ being the vector of all hyperparameters excluding $\boldsymbol{\sigma}$.

Updating the observed state can be performed in closed form by means of reformulating the nonlinear mapping (4.1) to

$$\boldsymbol{y}_i = g(x_i) + \boldsymbol{\sigma} \cdot \boldsymbol{v} \quad , \quad \boldsymbol{v} \sim \mathcal{N}(0,1) \, , \tag{4.44}$$

with $\boldsymbol{v}$ being uncorrelated with $\boldsymbol{\sigma}$. For a deterministic noise standard deviation σ, the model (4.44) is equivalent to (4.1) since $\boldsymbol{w} = \sigma \cdot \boldsymbol{v}$ with identical mean and variance. Here, the standard deviation $\boldsymbol{\sigma}$ of the observation noise is made explicitly accessible. This simplifies the update, as the mean $\underline{\mu}_k^y$ and covariance $\mathbf{C}_k^y$ of the observations as well as the cross-covariance $\mathbf{C}_k^{oy}$ between observed state and observations can be calculated exactly in closed form as shown in Paper K.

Assuming that the observed state $\underline{\boldsymbol{\xi}}_k^o$ and the observations $\underline{\boldsymbol{y}}_k$ are jointly Gaussian distributed, updating the observed state can be performed according to (2.9), which yields the desired conditional density $f\left(\underline{\xi}_k^o \middle| \underline{y}_{1:k}\right) \approx \mathcal{N}\left(\underline{\mu}_k^e, \mathbf{C}_k^e\right)$ with mean $\underline{\mu}_k^e$ and covariance $\mathbf{C}_k^e$ according to (2.9).

By means of the updated observed state it is now possible to update the unobserved part resulting in the Gaussian density $f\left(\underline{\xi}_k^u | \underline{\xi}_k^o\right) = \mathcal{N}\left(\underline{\mu}_k^u, \mathbf{C}_k^u\right)$

with mean $\underline{\mu}_k^u$ and covariance $\mathbf{C}_k^u$ according to (2.43) and (2.44), respectively.

To finalize the update step and thus, to obtain the desired joint posterior $f_k = \mathcal{N}\left(\underline{\xi}_k; \underline{\mu}_k^\xi, \mathbf{C}_k^\xi\right)$ with updated basis vectors and hyperparameters, the above results are combined according to

$$\underline{\mu}_k^\xi = \begin{bmatrix} \underline{\mu}_k^u \\ \underline{h}^\mathrm{T} \cdot \underline{\mu}_k^e \end{bmatrix}, \quad \mathbf{C}_k^\xi = \begin{bmatrix} \mathbf{C}_k^u & \mathbf{L}_k \mathbf{C}_k^e \underline{h} \\ \underline{h}^\mathrm{T} \mathbf{C}_k^e \mathbf{L}_k^\mathrm{T} & \underline{h}^\mathrm{T} \mathbf{C}_k^e \underline{h} \end{bmatrix} \tag{4.45}$$

with $\mathbf{L}_k \triangleq \mathbf{C}_k^{uo} \left(\mathbf{C}_k^o\right)^{-1}$ and $\underline{h}^\mathrm{T} \triangleq [1,0,0,\ldots,0]$. The first row in (4.45) corresponds to marginalizing out $\underline{\mathbf{g}}_k$.

Example 22: Synthetic Data

The performance of the RGP and the simultaneous regression and hyperparameter learning approach (denoted as RGP* in the following) is compared with a full GP as well as with the sparse GP methods SOGP and SPGP. For this purpose, data generated by means of two different synthetic functions are considered. The first function

$$\boldsymbol{y} = \tfrac{\boldsymbol{x}}{2} + \tfrac{25 \cdot \boldsymbol{x}}{1+\boldsymbol{x}^2} \cdot \cos(\boldsymbol{x}) + \boldsymbol{w} \quad , \quad \boldsymbol{w} \sim \mathcal{N}(0, 0.1) \tag{4.46}$$

is smooth but non-stationary. It is similar to the system model in (4.23). At each step k, 40 input-observation pairs are selected randomly from the interval $[-10, 10]$. In total 100 steps are performed. The active sets (SOGP, SPGP) and basis vectors (RGP, RGP*) comprise 50 elements, which are placed equidistant on the interval $[-10, 10]$. As second function

$$\boldsymbol{y} = \mathcal{N}(0.6, 0.04) + \mathcal{N}(0.15, 0.0015) + 4 \cdot H(0.3) + \boldsymbol{w} \,, \tag{4.47}$$

is considered, where $H(.)$ is the Heaviside step function with $H(a) = 0$ if $x \leq a$ and $H(a) = 1$ if $x > a$. This function has a discontinuity at $x = 0.3$ and was considered as a benchmark in [213]. The noise $\boldsymbol{w}$

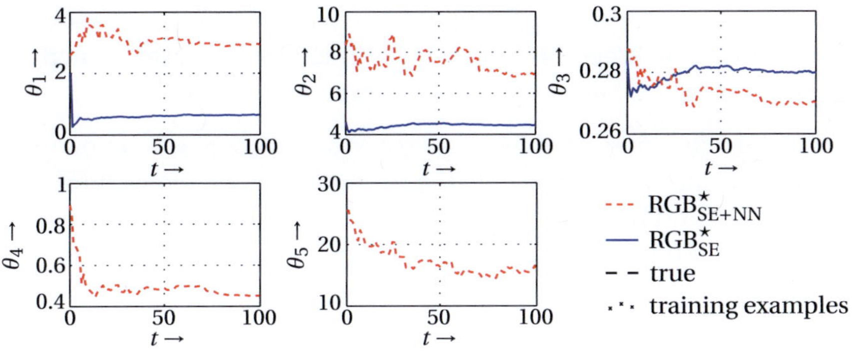

(a) Evolution of the hyperparameters of RGP* with SE (blue, solid) and SE+NN covariance function (red, dashed), where $\theta_1 = l_1$ (length-scale), $\theta_2 = \alpha_1$ (signal standard deviation), $\theta_3 = \sigma$ (noise standard deviation) are the parameters of the SE kernel and $\theta_4 = l_2$ (length-scale), $\theta_5 = \beta$ (signal standard deviation) are the parameters of the NN kernel.

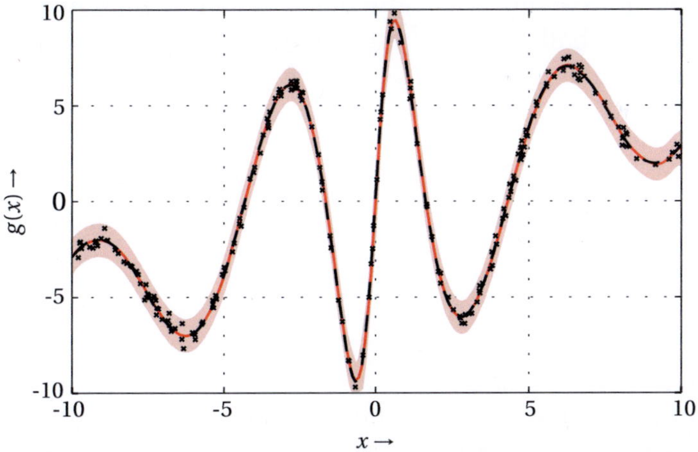

(b) True function (black, dashed line), regression result by RGP* with SE+NN covariance function together with 99% confidence area, and the training examples of step $t = 100$ (black crosses).

Figure 4.5: Exemplary regression result of proposed approach for function (4.46).

has variance $\sigma^2 = 0.16$. A total of 70 steps are performed, with 50 data points per step drawn from $[-2, 2]$. On this interval, 30 active set elements and basis vectors are placed equidistant.

The mean function of all GPs is zero and as covariance function two different ones are employed: the SE function (4.8) as well as the sum of SE and NN function (see (4.10)), denoted as SE+NN in the following. The hyperparameters for the full GP are optimized via evidence maximization (4.12). These optimized hyperparameters are also used for the RGP.

In Figure 4.5a on the previous page, an exemplary regression result of RGP* with SE+NN covariance function is depicted. The true function is accurately reconstructed. As shown in Figure 4.5b, the hyperparameters are adjusted over time and converge. This leads to improved regression results compared to the other hyperparameter learning approaches SOGP and SPGP as can be seen in Table 4.3. This holds for both covariance functions, whereas SE+NN yields better results as it is possible to capture the non-stationarity thanks to the non-stationary NN kernel. Compared to a full GP, RGP* is slightly inferior. The off-line hyperparameter optimization (4.12) provides optimal results and RGP* cannot improve further. With the optimal hyperparameters however, RGP performs close to a full GP but with significantly lower runtime.

The results in Table 4.4 indicate that off-line hyperparameter optimization (4.12) is not always optimal. Here, the hyperparameters learned by RGP* result in better estimates compared to all other algorithms with at the same time lower computational load. It is worth mentioning that RGP* is the only sparse approach that really exploits the properties of the SE+NN kernel resulting in an improved regression compared to the SE kernel.

Table 4.3: Average rmse, nll, and runtime for function (4.46).

	SE			SE+NN		
	rmse	nll	time in s	rmse	nll	time in s
Full GP	0.31 ± 0.02	0.25 ± 0.05	0.82	0.30 ± 0.02	0.24 ± 0.06	1.46
RGP	0.31 ± 0.02	0.26 ± 0.06	0.16	0.31 ± 0.03	0.24 ± 0.06	0.11
RGP*	0.37 ± 0.02	0.41 ± 0.14	0.65	0.35 ± 0.05	0.34 ± 0.12	1.81
SOGP	1.18 ± 0.03	7.49 ± 0.4	0.93	0.44 ± 0.03	0.80 ± 0.13	2.58
SPGP	0.54 ± 0.02	1.15 ± 0.11	13.05	0.39 ± 0.02	0.51 ± 0.09	24.40

Table 4.4: Average rmse, nll, and runtime for function (4.47).

	SE			SE+NN		
	rmse	nll	time in s	rmse	nll	time in s
Full GP	1.38 ± 0.41	1.70 ± 0.33	0.92	1.04 ± 0.43	1.41 ± 0.35	2.57
RGP	1.40 ± 0.38	1.98 ± 0.40	0.45	1.12 ± 0.35	1.86 ± 0.19	0.48
RGP*	0.98 ± 0.11	1.48 ± 0.23	0.38	0.88 ± 0.10	1.39 ± 0.15	1.14
SOGP	1.68 ± 0.07	2.89 ± 0.17	0.8	1.68 ± 0.07	2.88 ± 0.17	2.21
SPGP	1.63 ± 0.10	1.91 ± 0.06	5.6	1.65 ± 0.07	1.93 ± 0.05	10.85

Discussion

Directly modeling some of the hyperparameters by means of a Gaussian may not be appropriate in some cases. For instance, the length-scale hyperparameters of the SE covariance function in (4.8) have to be positive. To account for such constraints, a standard trick in GP regression is to transform the hyperparameters first and then to train the transformed parameters. After training, the inverse transformation is applied in order to obtain the original hyperparameters. In case of positive hyperparameters, the logarithm for transforming and the exponential function as inverse transformation are common. RGP* can directly be used to also train/estimate transformed hyperparameters.

Assuming Gaussian noise $\boldsymbol{w}$ in (4.1) is not reasonable for every application. Capturing a non-Gaussian distribution by the proposed methods can for instance be achieved via warping as proposed in [178]. Alternatively, f_k could be represented by means of a Gaussian mixture allowing for the application of techniques proposed in Section 3.4.

The computation and memory costs of RGP* for a single time step k scale with $\mathcal{O}\left(s \cdot n_k \cdot (m + n_\theta)^2 + n_k^3\right)$ and with $\mathcal{O}\left((m + n_\theta)^2\right)$, respectively, where s is the number of samples of the employed LRKF, n_k is the number of observations at step k, m is the number of basis vectors, and n_θ is the dimension of $\underline{\theta}$. If at each step k the same number of observations is processed, than the computational and memory costs are constant for each step for both RGP and RGP*. Furthermore and in contrast to a full GP the computational and memory costs do not increase over time, i.e., when more and more observations become available.

4.6 Summary

Assuming that no analytical system and measurement models are available, but GP representations of these models exist, the contributions made in this chapter are concerned with performing analytic filtering and smoothing as well as on-line learning of GP models:

- *Analytic filtering and smoothing for GP models:* For particular covariance functions, Gaussian filtering and smoothing for GP system and measurement models can be performed in closed-form, without the need of sampling or linearization. Given a sufficient amount of training data, filtering and smoothing are superior compared to many other Gaussian filters operating on the analytical models.
- *Recursive GP regression:* Especially for streaming data, there is a lack of sparse GP regression approaches in the state-of-the-art. The proposed RGP allows regression with constant computational and memory demand. This approach makes no restriction on the used mean and covariance functions.
- *On-line hyperparameter learning:* RGP can be extended in such a way that on-line hyperparameter learning for streaming data is possible. Here, updating the basis vectors and hyperparameters with new data is treated as a joint Gaussian filtering problem, which leads to a computationally efficient and accurate GP regression approximation.

Generally, there are many machine learning techniques for learning system and measurement models from data. GP regression however forms a Bayesian approach of this task, with close relationship to Gaussian filtering. This relationship is the main purpose for allowing the above contributions, where it is possible to benefit from the rich theoretical and algorithmic foundation of Gaussian filtering introduced and extended in Chapter 2.

5

Applications

For every major distribution and filtering group introduced in the previous three chapters a dedicated real-world application is studied in this chapter. These applications are:

- *Range-based localization:* Estimating the position and orientation of a moving object via Gaussian filtering (summarizes Papers L and M).
- *Gas dispersion source estimation:* Determining the location and strength of a gas release into atmosphere using Gaussian mixture filtering (Paper N).
- *Active Object recognition:* Effectively utilizing a movable camera for fast object recognition based on Gaussian process regression (Paper O).

This list already indicates that Bayesian filtering in general and the proposed solutions in particular are applicable in a broad range of real-world estimation problems.

5.1 Range-based Localization

In applications such as car navigation, mobile robot navigation, or telepresence, the position of a moving object is often localized based on range/ distance measurements between the object and known landmarks. These ranges can for example be measured by times of arrival or field strengths [169].

Existing range-based localization algorithms can be divided into two classes. Approaches of the first class assume exact (or almost exact) range measurements. As long as this assumption is satisfied, closed-form localization approaches as those in [15, 34, 43, 78, 119, 196], gradient descent algorithms [153], or methods based on linearization via Taylor-series expansion [66] perform very well. However, these approaches merely allow for a *static localization*, i.e., at every time step an independent location estimation is performed. Furthermore, accurate range measurements require specialized and expensive hardware.

Dealing with inaccurate measurements that may arise for example from signal strength information or ultrasonic range finders requires range-based localization approaches from the second class. Based on probabilistic models that capture measurement uncertainties—for instance arising from measurement noise or modeling errors—the object's position and velocity can be estimated by means of a Bayesian filter in a recursive fashion. This allows for *dynamic localization*, i.e., the combination of dead reckoning and static localization, for a smoother and more robust localization. The maybe most prominent range-based localization algorithm based on Bayesian filtering is used in GPS.

Example 23: GPS

The *global positioning system* (GPS) is the most widely used satellite-based navigation system. The localization principle employed in GPS is based on *multilateration*, i.e., measuring the distance between the object's position and several reference points or *landmarks* in

3D. In GPS the necessary distances are determined by using *time of arrival*. Here, the satellites send a signal to the receiver (the object) that includes information about the exact time of its broadcast. If the time of the receiver is synchronized with the system time of the satellites, which is the same for all satellites, the receiver is now capable of calculating the duration of signal transmission. By multiplying the duration of the transmission with the speed of light the required distance is obtained. The receiver typically uses an EKF for estimating its position and velocity based on the calculated distances and the dynamics of the object [99].

When assuming that the object can be considered as a point in space, the quantities of interest are merely position, velocity and sometimes acceleration. GPS for instance makes this assumption. In applications like telepresence however, where the extent of the object is important, also the orientation and corresponding angular velocities in 3D have to be estimated. For both problems—position *and* pose estimation—dynamic range-based localization algorithms utilizing Gaussian filtering are introduced in the following.

5.1.1 Position Estimation

At first it is assumed that the object can be considered a point as depicted in Figure 5.1a on the next page. In this case, the state $\underline{\boldsymbol{x}}^{\mathrm{T}} \triangleq \left[\underline{\boldsymbol{t}}^{\mathrm{T}}\ \dot{\underline{\boldsymbol{t}}}^{\mathrm{T}}\right]$ of the object consists of its position $\underline{\boldsymbol{t}} = [x\ y\ z]^{\mathrm{T}}$ and velocity $\dot{\underline{\boldsymbol{t}}} = [\dot{x}\ \dot{y}\ \dot{z}]^{\mathrm{T}}$.

Dynamic and Measurement Model

The dynamic behavior—the motion—of the object is described by means of the linear discrete-time dynamic system

$$\underline{\boldsymbol{x}}_{k+1} = \mathbf{A} \cdot \underline{\boldsymbol{x}}_k + \underline{\boldsymbol{w}}_k , \tag{5.1}$$

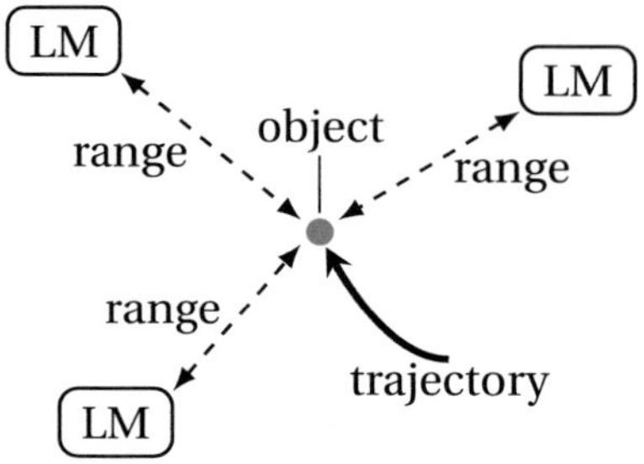

(a) Position estimation.

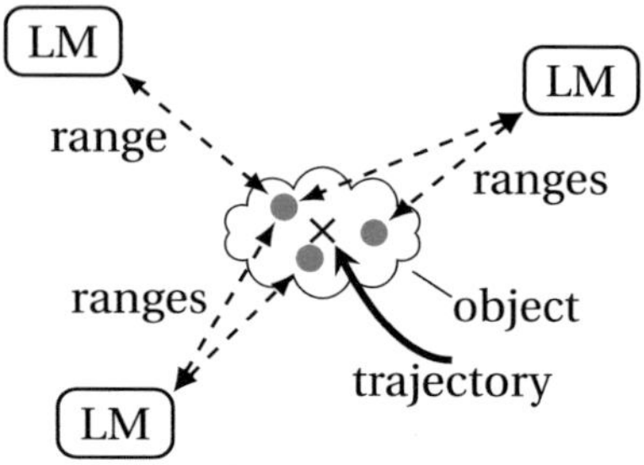

(b) Pose estimation.

Figure 5.1: Examples for range-based localization problems. Based on the measured ranges between landmarks (LMs) at known positions and the unknown position of the object, the object's trajectory has to be estimated. In case of pose estimation (b), the object possesses itself several landmarks (gray circles) to which range measurements can be performed. The center of mass is indicated by the cross.

where the noise $\underline{\boldsymbol{w}}_k$ is assumed to be zero-mean white Gaussian. For a *position velocity model* [207], the matrix $\mathbf{A}$ and the covariance of the process noise $\mathbf{C}^w$ are given by

$$\mathbf{A} = \begin{bmatrix} \mathbf{I} & T \cdot \mathbf{I} \\ \mathbf{0} & \mathbf{I} \end{bmatrix}, \quad \mathbf{C}^w = \begin{bmatrix} \frac{T^3}{3} \mathbf{C}_c^w & \frac{T^2}{2} \mathbf{C}_c^w \\ \frac{T^2}{2} \mathbf{C}_c^w & T \cdot \mathbf{C}_c^w \end{bmatrix}, \tag{5.2}$$

respectively, where T is the sampling time. $\mathbf{C}_c^w \triangleq \operatorname{diag}\left(\left[\sigma_{c,x}^2 \quad \sigma_{c,y}^2 \quad \sigma_{c,z}^2\right]\right)$ corresponds to the process noise covariance of the continuous time system model with $\sigma_{c,\xi}^2$ being the variances of dimension $\xi \in \{x, y, z\}$.

Range measurements to N landmarks at the known positions $\underline{S}_i \in \mathbb{R}^3$ with $i = 1, \ldots, N$ are incorporated. The nonlinear relation between the object position and the landmark position is given by

$$\boldsymbol{\rho}_{k,i} = \left\| \underline{S}_i - \underline{\boldsymbol{t}}_k \right\|_2, \tag{5.3}$$

where $\boldsymbol{\rho}_{k,i}$ is the Euclidean distance between object and landmark.

In a real scenario, the ranges cannot be measured exactly, i.e., measurement uncertainty has to be considered, which is usually done by incorporating a noise process into (5.3). Two possibilities arise for incorporation. In the first case, which is the *standard model*

$$\boldsymbol{\rho}_{k,i} = \left\| \underline{S}_i - \underline{\boldsymbol{t}}_k \right\|_2 + \underline{\boldsymbol{v}}_{k,i} \,, \tag{5.4}$$

the noise process $\boldsymbol{v}_{k,i}$ directly affects the range $\boldsymbol{r}_{k,i}$. In the second case

$$\boldsymbol{\rho}_{k,i} = \left\| \underline{S}_i - \underline{\boldsymbol{t}}_k - \underline{\boldsymbol{v}}_{k,i} \right\|_2 \,, \tag{5.5}$$

which is called *noise before non-linearity* [44], the noise process affects the difference between object and landmark position. This measurement model can be interpreted such that the positions of the landmarks are uncertain. In both measurement models, the noise process is assumed to be zero-mean white Gaussian. In the following, the second model (5.5) is considered for mainly two reasons: First, the standard model (5.4) is only appropriate in situations where the distance $\boldsymbol{\rho}_{k,i}$ is large compared to the variance of the noise $\boldsymbol{v}_{k,i}$. Otherwise, negative ranges are possible, which is not true in reality. This problem cannot occur in the second measurement model. Second, the model in (5.5) allows analytic moment matching.

Analytic Moment Matching

Thanks to the linearity of the dynamic model (5.2), the standard Kalman filter prediction step (2.12) can be employed for propagating the state from time step k to time step $k+1$. To obtain also analytic expressions for the moment integrals (2.8) in case of the measurement update, the measurement equation (5.5) is squared, which yields

$$\boldsymbol{d}_i \triangleq (\boldsymbol{\rho}_i)^2 = (\underline{S}_i - \underline{\boldsymbol{t}})^{\mathrm{T}} \cdot (\underline{S}_i - \underline{\boldsymbol{t}}) - 2 \cdot (\underline{S}_i - \underline{\boldsymbol{t}})^{\mathrm{T}} \cdot \underline{\boldsymbol{v}}_i + \underline{\boldsymbol{v}}_i^{\mathrm{T}} \cdot \underline{\boldsymbol{v}}_i \,, \tag{5.6}$$

where $\boldsymbol{d}_i$ is a squared range assumed to be calculated by $\hat{d}_i = \hat{\rho}_i^2$. Thus, the modified measurement equation (5.6) can be described in short term via

$$\boldsymbol{d}_i = h_i\left(\underline{\boldsymbol{t}}, \underline{\boldsymbol{v}}_i\right) \tag{5.7}$$

for a single measurement to landmark i and via

$$\underline{\boldsymbol{d}} = \underline{h}\left(\underline{\boldsymbol{t}}, \underline{\boldsymbol{v}}\right) \tag{5.8}$$

for measurements to all landmarks, where d_i and $h_i(.,.)$ are the ith element of $\underline{\boldsymbol{d}}$ and $\underline{h}(.,.)$, respectively. Hence, the vector of squared ranges $\hat{\underline{d}}$ is calculated according to $\hat{\underline{d}} = \hat{\underline{r}} \odot \hat{\underline{r}}$. The measurement noise $\underline{\boldsymbol{v}}$ in (5.8) is zero-mean with covariance matrix

$$\mathbf{C}^v = \begin{bmatrix} \mathbf{C}_1^v & \cdots & \mathbf{C}_{1,j}^v & \cdots & \mathbf{C}_{1,N}^v \\ \vdots & \vdots & \vdots & \vdots & \vdots \\ \mathbf{C}_{i,1}^v & \cdots & \mathbf{C}_{i,j}^v & \cdots & \mathbf{C}_{i,N}^v \\ \vdots & \vdots & \vdots & \vdots & \vdots \\ \mathbf{C}_{N,1}^v & \cdots & \mathbf{C}_{N,j}^v & \cdots & \mathbf{C}_N^v \end{bmatrix},$$

where $\mathbf{C}_{i,j}^v$ is the 3×3 covariance matrix between the ith and jth landmark. By assuming correlations between landmarks, an algorithm valid for many real-world application can be derived. The case of uncorrelated landmarks is a special case of the algorithm.

Due to the consideration of squared ranges, the measurement model in (5.6) is a polynomial of order two allowing closed-form calculations of the moment integrals (2.8). Hence, according to (2.9) the mean vector and covariance matrix of the posterior state estimate $\underline{\boldsymbol{x}}_k \sim \mathcal{N}\left(\underline{\mu}_k^e, \mathbf{C}_k^e\right)$ are calculated via

$$\begin{aligned} \underline{\mu}_k^e &= \underline{\mu}_k^p + \mathbf{C}_k^{xd} \cdot \left(\mathbf{C}_k^d\right)^{-1} \cdot \left(\hat{\underline{d}}_k - \underline{\mu}_k^d\right), \\ \mathbf{C}_k^e &= \mathbf{C}_k^p - \mathbf{C}_k^{xd} \cdot \left(\mathbf{C}_k^d\right)^{-1} \cdot \left(\mathbf{C}_k^{xd}\right)^{\mathrm{T}}, \end{aligned} \tag{5.9}$$

respectively, where $\underline{\mu}_k^d$ (squared measurement mean), $\mathbf{C}_k^d$ (squared measurement covariance matrix), and $\mathbf{C}_k^{xd}$ (cross-covariance between state and squared measurement) are given by[1]

$$\begin{aligned}
\underline{\mu}_k^d &= (\mathbf{V} \odot \mathbf{V})^{\mathrm{T}} \cdot \underline{1}_M + \underline{1}_N \cdot \mathrm{Tr}\left(\mathbf{P} \cdot \mathbf{C}_k^p \cdot \mathbf{P}^{\mathrm{T}}\right) + \mathbf{O}^{\mathrm{T}} \cdot \mathrm{diag}\left(\mathbf{C}^v\right) , \\
\mathbf{C}_k^d &= \mathbf{O}^{\mathrm{T}} \cdot \left(4 \cdot \left(\mathrm{vec}(\mathbf{V}) \cdot \mathrm{vec}(\mathbf{V})^{\mathrm{T}}\right) \odot \mathbf{T} + 2 \cdot \mathbf{T} \odot \mathbf{T}\right) \cdot \mathbf{O} , \\
\mathbf{C}_k^{xd} &= -2 \cdot \mathbf{C}_k^p \cdot \mathbf{P}^{\mathrm{T}} \cdot \mathbf{V} ,
\end{aligned} \tag{5.10}$$

respectively, with

$$\begin{aligned}
\mathbf{S} &\triangleq \begin{bmatrix} \underline{S}_1 & \dots & \underline{S}_N \end{bmatrix} , \\
\mathbf{P} &\triangleq \begin{bmatrix} \mathbf{I}_M & \mathbf{0}_M \end{bmatrix} , \\
\mathbf{V} &\triangleq \mathbf{S} - \left(\underline{1}_N\right)^{\mathrm{T}} \otimes \left(\mathbf{P} \cdot \underline{\mu}_k^p\right) , \\
\mathbf{O} &\triangleq \mathbf{I}_N \otimes \underline{1}_M , \\
\mathbf{T} &\triangleq \mathbf{C}^v + \mathbf{1}_N \otimes \left(\mathbf{P} \cdot \mathbf{C}^p \cdot \mathbf{P}^{\mathrm{T}}\right) ,
\end{aligned}$$

where vec(**V**) is the vectorized version of the matrix **V**, $\underline{1}_N$ is a vector of ones of dimension N, and $\mathbf{1}_N$ is a one matrix. The variable $M = 3$ stands for the three-dimensional space.

Example 24: Four Landmarks

The proposed analytical moment calculation (AMC) is compared against the EKF and UKF. For this purpose four landmarks with positions

$$\mathbf{S} = \begin{bmatrix} \underline{S}_1 & \dots & \underline{S}_4 \end{bmatrix} = \begin{bmatrix} -2 & -2 & 2 & 2 \\ -2 & 2 & -2 & 2 \\ 0 & 0 & 0 & 2 \end{bmatrix} \mathrm{m}$$

1 A detailed derivation can be found in Paper L.

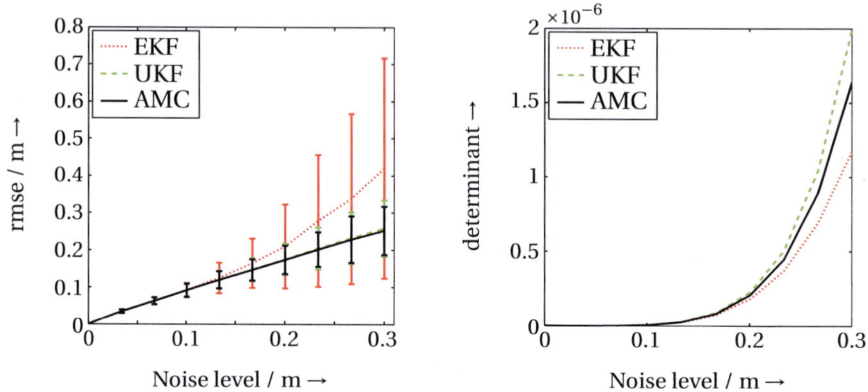

(a) The average rmse and its standard deviation.

(b) Mean of the determinant of the position covariance $\mathbf{C}^e_{k,t}$.

Figure 5.2: Result of the three estimators AMC, UKF, and EKF for different noise levels.

are considered. The measurement noise covariance matrix of each landmark i is assumed to be $\mathbf{C}^v_i = \mathbf{I} \cdot \sigma^2_n$ with $\sigma_n = {}^{(n-1)}/_3\,\mathrm{m}$ where $n = 1 \ldots 10$, i.e., ten different noise levels are investigated. For each noise level 1000 MC runs are simulated, where each run consists of 100 measurement steps.

The initial state at time step $k = 0$ has zero mean and covariance $\mathbf{C}_0 = 10 \cdot \mathbf{I}_6$. The sampling time is $T = 0.1\mathrm{s}$. The process noise covariance matrix comprises the elements $\sigma^2_{c,x} = \sigma^2_{c,y} = 0.01$, and $\sigma^2_{c,z} = 0.0001$.

In Figure 5.2a, the average rmse is depicted. For small noise, all three filters perform similar. If the noise increases, the rmse of the EKF increases much stronger compared to the other two filters. For a high noise level, the UKF and the proposed approach present comparable results, where the average rmses and the standard deviations of the AMC are slightly smaller.

The average determinant of the covariance matrix of the position estimate $\underline{t}_k$ of all test runs is shown in Figure 5.2b. Due to the linearization based on first-order Taylor-series expansions, the determinant of the EKF is too small and thus the EKF is too certain about its estimate. Hence, the estimation results are inconsistent, which is often a problem when using an EKF. On the other hand, LRKFs or analytic approaches as the AMC overcome this problem. The determinant of the AMC is smaller compared to the determinant of the UKF. Furthermore, as described before, the rmse of the AMC is smaller as well. All together, the AMC is more informative compared to the UKF.

The computational complexity of the above closed-form measurement update is in $\mathcal{O}(M^3 + M^2 \cdot N)$ for the mean $\underline{\mu}_k^d$, in $\mathcal{O}(M^2 \cdot N^3)$ for the covariance $\mathbf{C}_k^d$, and $\mathcal{O}(M^2 \cdot N)$ for the cross-covariance $\mathbf{C}_k^{xd}$, where typically $M \ll N$. For calculating the required moments in (5.10), only vector-matrix products and no additional matrix inversions or roots are required. For comparison, already the computational complexity of calculating the matrix square root required for an LRKF is in $\mathcal{O}(N^3 \cdot M^3)$.

5.1.2 Position and Orientation Estimation

In the following, the previous localization problem is generalized in order to allow the estimation of the pose—position and orientation—of an extended object in 3D. Hence, the object state is given by

$$\underline{x} \triangleq \begin{bmatrix} \underline{t}^{\mathrm{T}} & \underline{\dot{t}}^{\mathrm{T}} & \underline{r}^{\mathrm{T}} & \underline{\omega}^{\mathrm{T}} \end{bmatrix}^{\mathrm{T}}$$

where $\underline{t}$ and $\underline{\dot{t}}$ are the position and velocity in 3D, respectively, and $\underline{\omega}$ is the angular velocity in 3D . The orientation vector $\underline{r} \in \mathbb{R}^3$ is a so-called *rotation vector* with norm $\|\underline{r}\| = \pi$. Alternatives for describing the orientation are quaternions [108] or Euler angles [130]. Quaternions however are not minimal as they consist of four elements and the unit quaternions constraint can lead to inaccurate state estimates. Euler angles suffer from singularities and are not intuitive in usage. Rotation vectors instead are a minimal state representation. A further advantage of rotation vectors is that the dynamic behavior can be described by means of a nonlinear differential equation [22].

Dynamic Model

As the state now also comprises orientation related quantities, the dynamic model consists of two separate motion models: one for the translation and the second for the rotation. The translation model coincides with (5.1) and (5.2), respectively. The temporal evolution of the rotation vector $\underline{\boldsymbol{r}}_k$ is described by means of a nonlinear equation [17, 22]

$$\underline{\boldsymbol{r}}_{k+1} = \underline{\boldsymbol{r}}_k + T \cdot \underbrace{\left(\mathbf{I} + 0.5 \cdot \mathbf{R}(\underline{\boldsymbol{r}}_k) + a(\underline{\boldsymbol{r}}_k) \cdot \mathbf{R}(\underline{\boldsymbol{r}}_k) \cdot \mathbf{R}(\underline{\boldsymbol{r}}_k)\right)}_{\triangleq \Lambda(\underline{\boldsymbol{r}}_k)} \cdot \underline{\boldsymbol{\omega}}_k \,, \tag{5.11}$$

with

$$a(\underline{\boldsymbol{r}}_k) \triangleq \frac{1 - 0.5 \cdot \|\underline{\boldsymbol{r}}_k\|}{\|\underline{\boldsymbol{r}}_k\|^2} \cdot \cot\left(\frac{\|\underline{\boldsymbol{r}}_k\|}{2}\right) .$$

The system model for the angular velocity $\underline{\boldsymbol{\omega}}_k$ is assumed to be

$$\underline{\boldsymbol{\omega}}_{k+1} = \underline{\boldsymbol{\omega}}_k + \underline{\boldsymbol{w}}_k^{\omega} \,, \tag{5.12}$$

where $\underline{\boldsymbol{w}}_k^{w}$ is the process noise that affects the angular velocity. The process noise has zero mean and is Gaussian distributed with covariance $\mathbf{C}^{\omega}$.

By combining (5.2), (5.11) and (5.12), the system model for the pose estimation scenario can be written as

$$\underline{\boldsymbol{x}}_{k+1} = \begin{bmatrix} \mathbf{A} & \mathbf{0} & \mathbf{0} \\ \mathbf{0} & \mathbf{I} & \Lambda(\underline{\boldsymbol{r}}_k) \\ \mathbf{0} & \mathbf{0} & \mathbf{I} \end{bmatrix} \cdot \underline{\boldsymbol{x}}_k + \underline{\boldsymbol{w}}_k \,, \tag{5.13}$$

where the covariance of the process noise $\underline{\boldsymbol{w}}_k$ comprises the covariances of the process noises from the translation model $\mathbf{C}^t$ and the rotation model $\mathbf{C}^\omega$ according to

$$\mathbf{C}^w = \begin{bmatrix} \mathbf{C}^t & \mathbf{0} & \mathbf{0} \\ \mathbf{0} & \mathbf{0} & \mathbf{0} \\ \mathbf{0} & \mathbf{0} & \mathbf{C}^\omega \end{bmatrix} .$$

Measurement Model

As depicted in Figure 5.1b, in range-based pose estimation the measured ranges depend on known landmarks located on the extended object and on known landmarks in a global coordinate system. Thus, the ranges depend on both the unknown translation *and* rotation of the object with respect to the global coordinate system.

Example 25: Telepresence

An example application where knowing the pose is of interest is *large-scale telepresence* [153]. Here, the pose of a human has to be tracked for steering a robot. For tracking the pose of the user or the pose of several body parts, several emitters are located at known positions in a global coordinate system. They are emitting signals that are received by several sensors attached to the user, e.g., at a head-mounted display (see Figure 5.3a) or at gloves (see Figure 5.3b). Based on the emitted and received signals, ranges between emitters and sensors can be determined.

The relationship between measured ranges, translation, and rotation is given by

$$\boldsymbol{\rho}_{k,ij} = \left\| \underline{L}_j - \mathbf{D}(\underline{\boldsymbol{r}}_k) \cdot \underline{M}_i - \underline{\boldsymbol{t}}_k - \underline{\boldsymbol{v}}_{k,ij} \right\|_2 , \tag{5.14}$$

which resembles the noise before non-linearity range measurement model as in (5.5). Here, $\underline{L}_j$ is the position of the jth landmark with respect to the

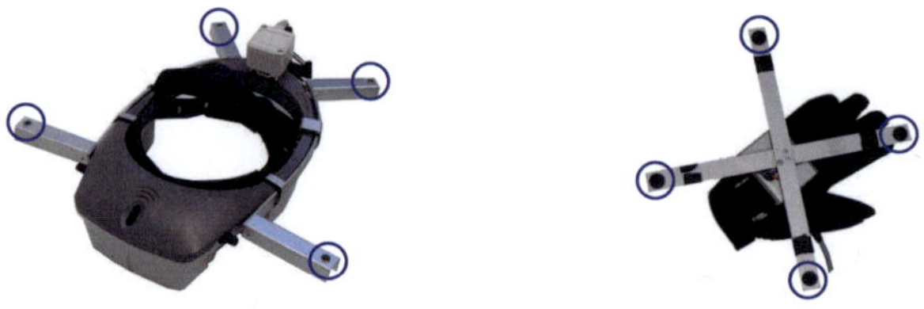

(a) Sensors at a head-mounted display.

(b) Sensors mounted at a glove.

Figure 5.3: Deployment of sensors (marked by blue circles) for tracking the pose of a user or of body parts in telepresence applications. Images taken from [111].

global coordinate system and $\underline{M}_i$ is the position of the ith landmark with respect to the object coordinate system. $\underline{\boldsymbol{v}}_{k,ij}$ is the measurement noise between landmark $\underline{L}_j$ and landmark $\underline{M}_i$. The term $\boldsymbol{\rho}_{k,ij}$ is the measured range between these two landmarks, while $\underline{\boldsymbol{d}}_k$ comprises all possible measurements between global and object landmarks. The term $\mathbf{D}(\cdot)$ is the rotation matrix parametrized by the rotation vector $\underline{\boldsymbol{r}}_k$

$$\mathbf{D}(\underline{\boldsymbol{r}}_k) = \mathbf{I} + \frac{\sin\left(\|\underline{\boldsymbol{r}}_k\|\right)}{\|\underline{\boldsymbol{r}}_k\|} \cdot \mathbf{R}(\underline{\boldsymbol{r}}_k) + \frac{1-\cos\left(\|\underline{\boldsymbol{r}}_k\|\right)}{\|\underline{\boldsymbol{r}}_k\|^2} \cdot \mathbf{R}(\underline{\boldsymbol{r}}_k) \cdot \mathbf{R}(\underline{\boldsymbol{r}}_k) ,$$

known as *Rodrigues formula,* with

$$\mathbf{R}(\underline{\boldsymbol{r}}_k) = \begin{bmatrix} 0 & -\boldsymbol{r}_z & \boldsymbol{r}_y \\ \boldsymbol{r}_z & 0 & -\boldsymbol{r}_x \\ -\boldsymbol{r}_y & \boldsymbol{r}_x & 0 \end{bmatrix}$$

being a skew-symmetric matrix.

Semi-Analytic Moment Matching

The system model (5.13) is conditionally linear, i.e., if the rotation vector is set to a fixed value, the system model becomes linear and the prediction for each value can be performed by using the well-known Kalman predictor equation.

To facilitate an analytical solution of the measurement update in closed from, the measurement equation (5.14) is squared as in Section 5.1.1, which yields

$$\underline{\boldsymbol{d}}_{k,ij} \triangleq \left(\boldsymbol{\rho}_{k,ij}\right)^2 = \left(\underline{g}_{ij}(\underline{\boldsymbol{r}}_k) - \underline{\boldsymbol{t}}_k - \underline{\boldsymbol{v}}_{k,ij}\right)^{\mathrm{T}} \cdot \left(\underline{g}_{ij}(\underline{\boldsymbol{r}}_k) - \underline{\boldsymbol{t}}_k - \underline{\boldsymbol{v}}_{k,ij}\right) , \quad (5.15)$$

with

$$\underline{g}_{ij}(\underline{\boldsymbol{r}}_k) \triangleq \underline{L}_j - \mathbf{D}(\underline{\boldsymbol{r}}_k) \cdot \underline{M}_i \, . \quad (5.16)$$

But in contrast to the previous section, squaring the measurement model alone is not sufficient due to the nonlinear term (5.16). By conditioning on $\underline{\boldsymbol{r}}$, however, (5.16) becomes affine and the entire measurement model becomes quadratic. Thus, the measurement model is conditionally integrable and the nonlinear-nonlinear decomposition proposed in Section 2.5.2 can be applied. The analytically integrable system state comprises $\underline{\boldsymbol{x}}_a^{\mathrm{T}} \triangleq \left[\underline{\boldsymbol{t}}^{\mathrm{T}} \quad \underline{\dot{\boldsymbol{t}}}^{\mathrm{T}} \quad \underline{\boldsymbol{w}}^{\mathrm{T}}\right]$, while the sampled state is $\underline{\boldsymbol{x}}_s \triangleq \underline{\boldsymbol{r}}$. For the latter, sampling via LRKFs can be employed. In doing so, for every fixed sample value of the rotation vector, the closed-form solutions in (5.10) for the moment integrals can be used with some minor modifications: the state dimension is now nine instead of six and the number of measured ranges is significantly higher as pair-wise measurements between object landmarks and global landmarks are incorporated.

Example 26: Two-dimensional Localization

A two-dimensional coordinate system is considered containing four sensors (global landmarks) and four emitters (object landmarks) with positions

$$\begin{bmatrix} \underline{M}_1^{\mathrm{T}} \\ \underline{M}_2^{\mathrm{T}} \\ \underline{M}_3^{\mathrm{T}} \\ \underline{M}_4^{\mathrm{T}} \end{bmatrix} = \begin{bmatrix} -0.2 & -0.2 \\ -0.2 & 0.2 \\ 0.2 & -0.2 \\ 0.2 & 0.2 \end{bmatrix} \mathrm{m}\,, \quad \begin{bmatrix} \underline{L}_1^{\mathrm{T}} \\ \underline{L}_2^{\mathrm{T}} \\ \underline{L}_3^{\mathrm{T}} \\ \underline{L}_4^{\mathrm{T}} \end{bmatrix} = \begin{bmatrix} -2 & -2 \\ -2 & 2 \\ 2 & -2 \\ 2 & 2 \end{bmatrix} \mathrm{m}$$

with respect to the object coordinate system and the global coordinate system, respectively. At different noise levels ranging from [0.000001,...,0.3] m, 1000 random trajectories are generated, where the sampling time was $T = 0.1\,\mathrm{s}$. The noise process is assumed as isotropic.

The proposed SAGF is compared to the UKF. For the UKF a decomposition into directly observed states $\underline{\boldsymbol{t}}$, $\underline{\boldsymbol{r}}$ and indirectly observed states $\underline{\dot{\boldsymbol{t}}}$, $\underline{\boldsymbol{\omega}}$ as proposed in Section 2.5.1 is used. Furthermore, due to the fact that the measurement noise is mapped through the nonlinear transformation, it has to be approximated with samples as well. In total 71 sample points are required for the UKF. On the other hand, the proposed approach only has to approximate the rotation by sample points.

For the 2D case, the system equation (5.13) becomes linear and thus, the Kalman filter prediction can be using directly. The entries of the continuous process noise covariance are set to be $\mathbf{C}_c^t = \mathrm{diag}([0.1\ 0.1])$ and $C_c^\omega = 0.1$. The initial state has zero mean and its covariance matrix comprises $\mathbf{C}_0^t = \mathrm{diag}([10\ 10])$, $\mathbf{C}_0^{\dot{t}} = \mathrm{diag}([10\ 10])$, $\sigma_{r,0}^2 = 0.001$, and $\sigma_{\omega,0}^2 = 0.0001$.

The estimation performance in terms of the rmse of both estimators is almost identical. Regarding the computational effort, the proposed approach only has to determine sample points for one dimension, which can be implemented very efficiently. On the other hand, the UKF calculates a matrix square root of the covariance of the noise and the directly observed state. This operations is computationally involved considering the size of the combined covariance matrix, which is 35×35. In the simulation, the proposed approach is *three times faster* than the standard approach.

5.2 Gas Dispersion Source Estimation

If a hazardous gas has been released—either accidentally or deliberately—into atmosphere, it is of paramount importance to gain knowledge of this event at an early stage in order to increase the effectiveness of counter measures for protecting the public and for mitigating the harmful effects. By means of so-called *atmospheric dispersion models* (ADMs), it is possible to predict the concentration spread of the released gas in space and time. These models, however, merely provide reliable predictions, if the characteristics of the gas source are known precisely. To determine or estimate the source characteristics it necessary to solve an *inverse problem*, where one has to infer the location and strength of the gas release from concentration measurements of spatially distributed sensors.

In general, solution methods of the source estimation problem can be classified into *forward* and *backward* methods [143]. Forward methods employ an forward-running ADM multiple times in order to find an estimate of the source that best describes the given concentration measurements. Here, the mostly used techniques are based on Bayesian inference in combination with Monte Carlo sampling. Sequential Monte Carlo methods as described in [179, 217] employ a set of samples or particles that forms the posterior probability distribution of the source parameters. This distribution is updated by means of Bayes' rule whenever new concentration measurements from sensors are available. In contrast to this online procedure, Markov chain Monte Carlo (MCMC) methods process all acquired concentration measurements in a batch in order to determine the posterior distribution. For this purpose, samples are drawn from the posterior distribution by simulating a Markov chain that has the desired posterior distribution as its stationary distribution. Given a properly constructed Markov chain it can be shown that MCMC reaches the stationary distribution after a typically large number of sampling steps, which is known as the burn-in phase. Application of MCMC to source estimation can be found for instance in [23, 80, 170].

Backward methods instead perform only one model run in the reverse direction from the sensors to the source. Commonly used techniques are backtracking, where an inverse version of an ADM is utilized (see e.g. [82]), and variational methods, where a cost function between model predictions and concentration measurements is optimized (see e.g. [155, 185]). The backward approach is preferred over forward methods, when the number of sources is larger than the number of sensors [143].

Most of the above state-of-the-art methods allow merely an off-line batch source estimation. The AGMF proposed in Section 3.4.2, however, facilitates an on-line estimation, where concentration measurements are processed continually. In doing so, timely information about the source location and strength can be provided allowing fast responses. For the AGMF, the so-called *Gaussian plume dispersion model* is utilized as a forward model, which facilitates predicting the gas dispersion in closed-form with low computational overhead.

5.2.1 Atmospheric Dispersion Models

In the following, $c(\underline{x}, t)$ is the concentration of the substance at position $\underline{x} = [x\ y\ z]^{\mathrm{T}} \in \mathbb{R}^3$ and at time $t \geq 0$. The concentration follows the *advection-diffusion equation*

$$\frac{\partial c(\underline{x}, t)}{\partial t} = \nabla \cdot \left(\mathbf{D} \cdot \nabla c(\underline{x}, t) - \underline{v} \cdot c(\underline{x}, t)\right) + s(\underline{x}, t) \tag{5.17}$$

with $\nabla \triangleq [\partial/\partial x\ \partial/\partial y\ \partial/\partial z]^{\mathrm{T}}$ (see e.g. [81]). The term $\mathbf{D} \cdot \nabla c(\underline{x}, t)$ describes the diffusion according to Fick's law with diffusion matrix $\mathbf{D}(\underline{x}, t)$ and the term $\underline{v} \cdot c(\underline{x}, t)$ represents linear advection due to wind with velocity $\underline{v}(\underline{x}, t)$. Finally, $s(\underline{x}, t)$ is a source or sink term.

Analytical solutions of (5.17), i.e., functions $c(\underline{x}, t)$ satisfying the equation, exist merely for some special cases. One such special case employed for source estimation in the following is the Gaussian plume dispersion model. In order to obtain a closed-form solution, the Gaussian plume model requires several assumptions:

1. The substance is emitted at a constant *rate* $q > 0$ from a single point source at location $\underline{x}_s \triangleq [0\ 0\ z_s]^{\mathrm{T}}$. Thus, the source term $s(\underline{x}, t)$ in (5.17) becomes

$$s(\underline{x}, t) = q \cdot \delta(x) \cdot \delta(y) \cdot \delta(z - z_s) \ .$$

2. The wind is constant with velocity $v \geq 0$ and the wind direction is along the x-axis. Hence, the velocity in (5.17) becomes $\underline{v} = [v\ 0\ 0]^{\mathrm{T}}$.
3. The diffusion is a function of the downwind distance (positive x-axis) only. Furthermore, it is assumed that the advection dominates the diffusion in wind direction. Thus, the diffusion along the x-axis can be neglected and $\mathbf{D} = \operatorname{diag}\left(\left[0\ K_y(x)\ K_y(x)\right]\right)$ with eddy diffusion coefficients K_y, K_z.
4. The terrain is flat and the ground cannot be penetrated by the substance.
5. The solution is steady state, i.e., time independent.

Based on these assumptions and additional boundary conditions that force vanishing concentrations at infinite distance from the source and at upwind distances, (5.17) has the time-invariant solution

$$c(\underline{x}) = \frac{q}{2\pi \cdot v \cdot \sigma_y \sigma_z} \cdot \exp\left(-\frac{y^2}{2\sigma_y^2}\right) \cdot \left[\exp\left(-\frac{(z-z_s)^2}{2\sigma_z^2}\right) + \exp\left(-\frac{(z+z_s)^2}{2\sigma_z^2}\right)\right] , \qquad (5.18)$$

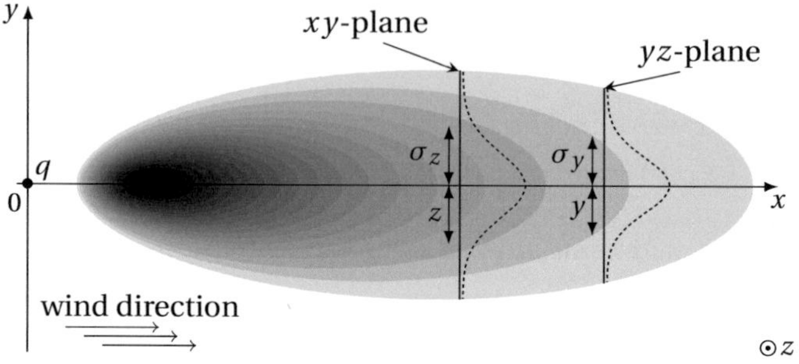

Figure 5.4: Ground level concentrations according to the Gaussian plume model, where black indicates the highest concentration level. The dotted lines indicate the profile of the plume when cutting through the plume in parallel with the xy-plane and yz-plane.

which is the well-known *Gaussian plume dispersion model* (for a detailed derivation see [185]). Here, σ_y and σ_z are the so-called standard deviations of the Gaussian concentration distribution. They depend on the stability of the atmosphere and are both functions of x. They can be obtained via integrating K_y and K_z in downwind direction or—more practically—can be determined via the Pasquill-Gifford stability classification scheme [38, 138].

Example 27: Gaussian Gas Plume

Consider a source emitting a gas contaminant from a height of $z_s = 4\,\mathrm{m}$. The wind velocity is $v = 2\,\mathrm{m/s}$. The atmospheric stability is of class D, which is "neutral" according to the Pasquill-Gifford classification. The corresponding ground level concentration is depicted in Figure 5.4. The gas plume spreads along the x-axis (downwind) and its shape is that of a Gaussian in planes normal to the x-axis.

The Gaussian plume model (5.18) is employed as it is widely used and suitable for describing short range substance releases. Furthermore, being an analytical model, it allows for an on-line and computationally lightweight estimation of the unknown parameters. It can be extended for arbitrary wind directions ϕ and arbitrary source location $\underline{x}_s = [x_s \; y_s \; z_s]^{\mathrm{T}}$ by means of straightforward translation and coordinate rotation, which yields the model

$$c(\underline{x}) = \frac{q}{2\pi \cdot v \cdot \sigma_y \sigma_z} \cdot \exp\left(-\frac{(1+2\sin(\phi)\cos(\phi)) \cdot (y-y_s)^2}{2\sigma_y^2}\right) \cdot \left[\exp\left(-\frac{(z-z_s)^2}{2\sigma_z^2}\right) + \exp\left(-\frac{(z+z_s)^2}{2\sigma_z^2}\right)\right] , \qquad (5.19)$$

where σ_y and σ_z are now functions of x_s, x, y, and ϕ. This model is employed in the following.

5.2.2 Parameter Estimation

Estimating the source rate q and source location $\underline{x}_s$ is a so-called *parameter estimation* problem. That is, the quantities of interest are time-invariant; there is no dynamical model driving the parameters over time. Instead, only data in form of concentration measurements from a set of spatially distributed sensors is available to determine the parameters. It is assumed that the measurements become available sequentially over time, i.e., batch or off-line estimation is impractical. Additional parameters like wind speed or direction are assumed to be known, as they can be provided reliably from external sources like weather stations.

In order to apply the AGMF, an appropriate measurement model is required that relates the concentration measurements to the unknown source parameters $\underline{\boldsymbol{\theta}}^{\mathrm{T}} \triangleq [\underline{\boldsymbol{x}}_s^{\mathrm{T}} \; \boldsymbol{q}]$. The Gaussian plume model reflects this relations. Thus, suppose that the measurement $\hat{z}_k$ is acquired by a sensor at location $\underline{x}_r \triangleq \left[x_r\ y_r\ z_r\right]^{\mathrm{T}}$ at time step k, the resulting measurement models is given by

$$\hat{z}_k = c\left(\underline{x}_r;\underline{\boldsymbol{\theta}}\right) + \boldsymbol{v}_k\,, \quad \boldsymbol{v}_k \sim \mathcal{N}\left(0,\sigma_v^2\right), \tag{5.20}$$

where $c\left(\underline{x}_r;\underline{\boldsymbol{\theta}}\right)$ is the true concentration value according to (5.19) and $\boldsymbol{v}_k$ is the sensor's noise, which is assumed to be independent in time and space. It is worth mentioning that in case of multiple sources, the concentration measurement can be written as

$$\hat{z}_k = \sum_{i=1}^{L} c_i\left(\underline{x}_r;\underline{\boldsymbol{\theta}}_i\right) + \boldsymbol{v}_k\,, \tag{5.21}$$

where the superposition of the concentration values $c_i\left(\underline{x}_r;\underline{\boldsymbol{\theta}}_i\right)$ of the sources $i = 1\ldots L$ is exploited (see [185]).

Example 28: Indianapolis Field Study

To demonstrate the performance of the AGMF in estimating the source parameters $\underline{\boldsymbol{\theta}}$, the data acquired during the EPRI Indianapolis field study is considered, where SF_6 tracer gas was released from a $z_s = 83.8\,\mathrm{m}$ stack at a power plant in Indianapolis, Indiana, USA. Data was recorded by 160 ground-level sensors over 19 days in September and October 1985 for 8 to 9 hours every day. Details about the field study and the data can be found in [79].

In Figure 5.5, the locations of the sensors and the sensors' concentration measurements of the 19th September 1985 are depicted. Even though all sensor measure at a hourly rate, the measurements are processed sequentially to demonstrate the on-line estimation capability of the AGMF.

The source is located at the origin and the emission rate of the tracer gas is $q = 0.0041\,\mathrm{g/s}$. Information about wind speed, wind direction, and atmospheric stability was made available by meteorological

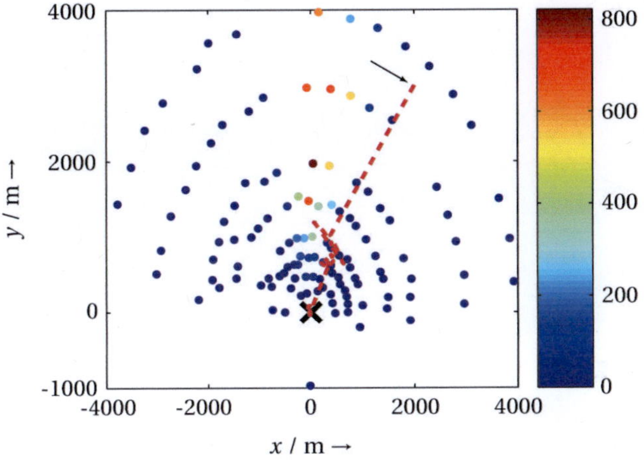

Figure 5.5: Trajectory of the estimated source location (red, dashed) and the true location of the source (black cross). Circular markers denote the sensor locations colored with the measured concentration in parts per trillion (ppt).

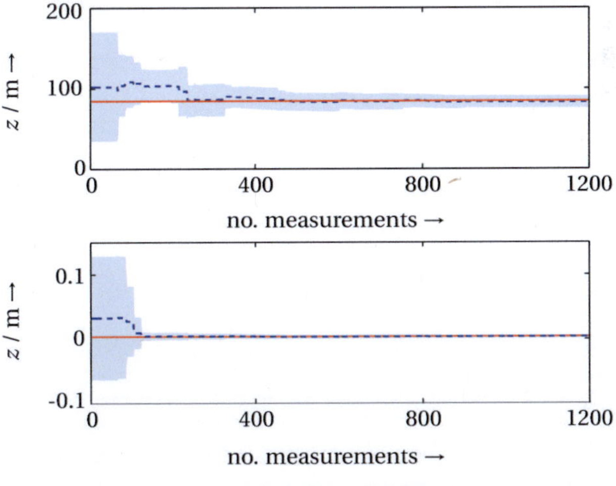

Figure 5.6: Estimate of source height z and emission rate q with increasing number of measurements. The shaded area denotes the 3-sigma confidence region and the red line indicates the true value.

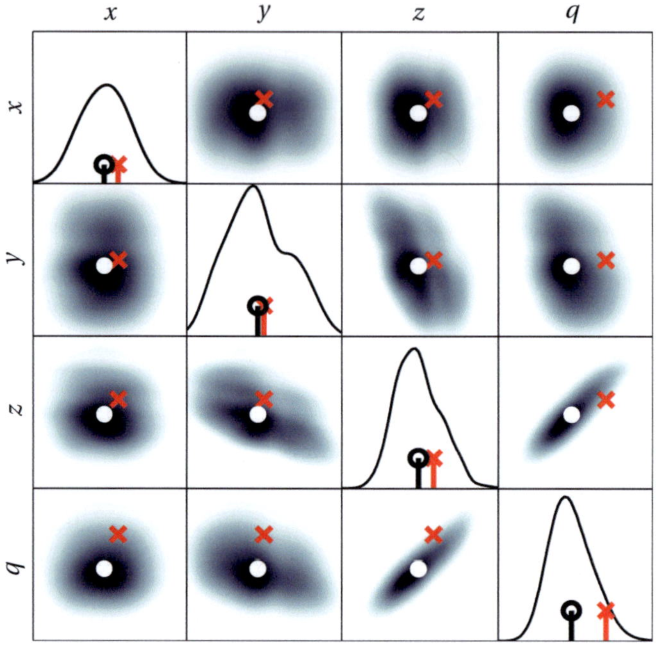

Figure 5.7: Bivariate posterior densities of the AGMF source estimate. The diagonal plots are the univariate marginal densities. Red crosses indicate the true value, while white and black circles, respectively, denote the mean of the respective density.

observations. The initial estimate of the source at time step $k = 0$ is given by a single Gaussian with mean vector and covariance matrix

$$\hat{\underline{\theta}}_0 = [2000\ 3000\ 102\ 0.033]^{\mathrm{T}}\,, \quad \mathbf{C}_0 = \mathrm{diag}\left(\left[10^6\ 10^6\ 500\ 0.001\right]\right)$$

respectively. Figure 5.5 and Figure 5.6 show the convergence of the source estimate towards the true source location over time and with increasing number of concentration measurements, respectively. It is important to note that many sensor measurements (typically 60%-70%) provide a concentration measurement of almost zero as most of the sensor are outside the gas plume, as can be seen in Figure 5.5. This explains the step-wise convergence of the estimate and reduction of the variance in Figure 5.6.

The posterior density $f_k^e(\underline{\theta})$ after all $k = 1200$ measurements is depicted in Figure 5.7. It can be seen that the mean of the estimate is close to the true source parameters. Slight deviation from the ground truth is only observed for the emission rate, but still the true parameters are within the high confidence region of the estimate. Thus, the AGMF is not overconfident.

5.3 Active Object Recognition

Research on computer vision mostly focuses on the object or scene observed by the camera system. It is assumed that the parameters of the camera (e.g., position, illumination, or focus) are given or determined off-line in a time-consuming trial-and-error process involving human interaction. Particular operations are then applied on the acquired images in order to solve the considered vision task like recognizing an object. In such *passive* vision systems, the camera parameters are not adapted on-line. This is in contrast to an *active* vision system, where the next camera observation is carefully planned based on the previously acquired images and prior information about the considered scene.

While various approaches for passive object recognition exist (see e.g. [190] and references therein), active object recognition still is in its early stages. One of the first approaches to active object recognition can be found in [21], where the object models are learned via the eigenspace approach introduced in [127]. The planning algorithm greedily chooses the view that leads to the maximum entropy reduction of the object hypotheses. In [56], from a finite set of views the one maximizing the mutual information between observations and classes is selected. The approach

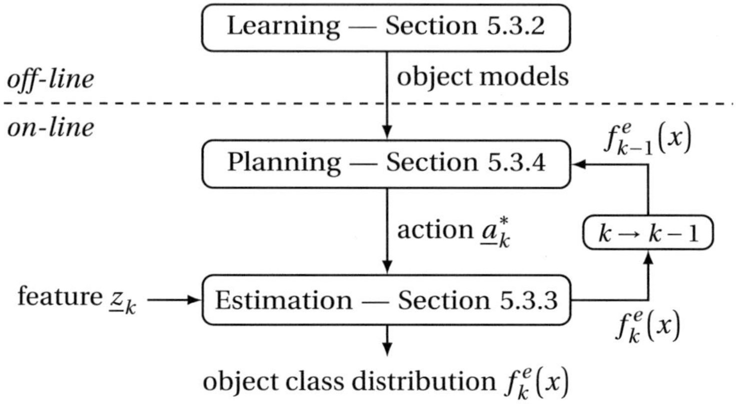

Figure 5.8: Flow chart of the active object recognition system.

is designed for arbitrary features, but requires approximate mutual information calculation via Monte Carlo sampling, which prevents a direct extension to continuous views. An upper bound of the Jeffrey divergence is employed in [112]. Again, merely a finite set of viewpoints is considered. Reinforcement learning approaches for active object recognition are proposed in [51, 135]. Here, learning the object models and planning is performed simultaneously. A comparison of some of the aforementioned approaches can be found in [50].

The active object recognition method proposed in this thesis consists of two parts as depicted in Figure 5.8. In the off-line *learning* part for each object a so-called object model is created. For varying camera parameters, e.g., focus or position, 2D images of each 3D object are generated. GP regression is then applied on the sample images to learn the object models.

In the on-line *recognition* part, planning the next-best camera view and Bayesian state estimation are performed alternately. For planning, mutual information is maximized with respect to the camera parameters. Based on the chosen parameter, the object estimate is updated via Bayesian estimation under consideration of the learned object models.

In contrast to prior art, the proposed method is very general as it is not restricted to specific image features. Furthermore, camera parameters can be arbitrary and continuous valued. All derivations hold for arbitrary GP covariance functions.

5.3.1 Object Classification

Object recognition can be considered a classification task where the state corresponds to the object class $x \in \mathcal{X} \triangleq \{x_1, x_2, \ldots x_N\} \subset \mathbb{N}$, with N being the finite number of possible object classes. For classification purposes, the state is represented by means of discrete random variable $\boldsymbol{x} \in \mathcal{X}$. Based on a feature vector $\underline{z}_k \in \mathcal{Z} \subseteq \mathbb{R}^{n_z}$ acquired from images at time/stage $k = 0, 1, \ldots$, the goal is to estimate the true latent object class. The measurement model

$$\underline{\boldsymbol{z}}_k = \underline{h}(\boldsymbol{x}, \underline{a}_k) + \underline{\boldsymbol{v}}_k \tag{5.22}$$

relates the feature vector with the object class. The quantity $\underline{a}_k \in \mathcal{A} \subseteq \mathbb{R}^{n_a}$ is the *camera parameter* that allows actively driving the classification process. Potential camera parameters are position, orientation, focal length, or exposure time, just to name a few. For active object recognition, an appropriate camera parameter has to be chosen at every stage k in order to improve the recognition performance and speed.

5.3.2 Learning

An analytical expression of the measurement model (5.22) is not available in general as it describes a complex transformation of a potentially high-dimensional feature vector to an abstract object class. To overcome this issue, a GP model is learned to represent (5.22). As the feature is typically multi-dimensional, for each dimension $e = 1 \ldots n_z$ of $\underline{\boldsymbol{z}}_k$ a separate GP is learned independently using the same training inputs $\mathbf{X}$ but different training outputs $\underline{\hat{z}}^e \triangleq \left[\hat{z}_1^e \ \ldots \ \hat{z}_n^e\right]^{\mathrm{T}}$. An alternative to this procedure are so-called *multi-output* GPs [25].

Furthermore, learning the GPs for each feature vector dimension has to be performed independently for each object class x_i, $i = 1\ldots N$. This results in N multi-variate GPs $\underline{\boldsymbol{h}}_i(.) \sim \mathcal{GP}$ of dimension n_z named *object models* in the following. To learn an object model $\underline{\boldsymbol{h}}_i$, samples $\underline{a}_l$, $l = 1\ldots n$ of the parameter space $\mathcal{A}$ are used as training inputs $\mathbf{X}$. For each input sample $\underline{a}_l$, an object of the class $x_i \in \mathcal{X}$ is observed by the camera resulting in the feature vector $\underline{\hat{z}}_l = \left[\hat{z}_l^1 \ \hat{z}_l^2 \ \ldots \ \hat{z}_l^{n_z}\right]^{\mathrm{T}}$ acting as training output. In total, for n_z output dimensions and N object classes, $n_z \times N$ GPs are learned. Since learning these measurement models is an off-line task (see Figure 5.8), the required computation time is independent of the computation time for object recognition. Furthermore, for high-dimensional features, which may be obtained for instance by means of the scale-invariant feature transform (SIFT, [116]), dimensionality reduction techniques like principal component analysis [2] or GP latent variable models [200] can be employed in order to reduce the number of GPs to be learned.

5.3.3 Estimation

To estimate the object class for a given camera parameter $\underline{a}_k$ and feature vector $\underline{\hat{z}}_k$ the Bayesian measurement update step (1.7) is employed[2], where the density $f_k^e(x)$ in the given object recognition task corresponds to a discrete distribution modeled as a mixture of Kronecker deltas according to

$$f_k^e(x) = \sum_{i=1}^{N} \omega_{k,i} \cdot \delta_{x,i} \ . \tag{5.23}$$

2 Due to the implicit assumption that the feature vectors $\underline{z}_k$ for $k = 0, 1, \ldots$ are conditionally independent given $\boldsymbol{x}$, this form of Bayesian classification/object recognition is known as *naive Bayes classifier* [61]. Even though this assumptions might not be true, naive Bayes classifiers showed a good classification performance in practice [58].

As the state is static—the object does not change its class over time—no prediction is performed and thus it holds that $f_k^p(x) \equiv f_{k-1}^e(x)$. The weight $\omega_{k,i}$ represents the probability that object x belongs to class i. The weights are non-negative and sum up to one. The measurement update boils down to updating the weights whenever a new feature vector $\hat{\underline{z}}_k$ is available. Before providing the weight update equation, it is first necessary to investigate the structure of the likelihood.

Likelihood

In contrast to (1.7), the likelihood in the considered recognition task also depends on the camera parameter. In case of a given object class $x = i$, the likelihood $f(\hat{\underline{z}}_k | x = i, \underline{a}_k)$ corresponds to the GP $\underline{\boldsymbol{h}}_i$. If in addition the camera parameter $\underline{a}_k$ is given, the likelihood becomes a Gaussian density $\mathcal{N}(\hat{\underline{z}}_k; \underline{\mu}_{k,i}^z, \mathbf{C}_{k,i}^z)$ with mean vector and covariance matrix according to

$$
\begin{aligned}
\underline{\mu}_{k,i}^z &= \left[\mu_{k,i}^1 \; \mu_{k,i}^2 \; \cdots \; \mu_{k,i}^{n_z}\right]^{\mathrm{T}} , \\
\mathbf{C}_{k,i}^z &= \operatorname{diag}\left(\left(\sigma_{k,i}^1\right)^2, \left(\sigma_{k,i}^2\right)^2, \ldots, \left(\sigma_{k,i}^{n_z}\right)^2\right) ,
\end{aligned}
\tag{5.24}
$$

respectively. The elements in (5.24) corresponding to dimension $e = 1 \ldots n_z$ are calculated according to (4.6) and (4.7), respectively, with $\underline{a}_k$ acting as test input $\underline{x}$ and $\hat{\underline{z}}^e$ being the training output vector $\underline{y}_{\mathcal{D}}$. Overall, the likelihood for a fixed $\underline{a}_k$ can be characterized by means of the conditional density

$$
f(\hat{\underline{z}}_k | x, \underline{a}_k) = \sum_{i=1}^{N} \delta_{x,i} \cdot \mathcal{N}\left(\hat{\underline{z}}_k; \underline{\mu}_{k,i}^z, \mathbf{C}_{k,i}^z\right) . \tag{5.25}
$$

It is important to note that for a fixed feature vector $\hat{\underline{z}}_k$—as required for solving Bayes' equation—the conditional density in (5.25) becomes a weighted sum of Kronecker deltas as in (5.23), because all Gaussian components are evaluated at $\hat{\underline{z}}_k$ and thus, become scalar weighting coefficients.

Posterior Weights

Given the likelihood in (5.25), the measurement update can be evaluated analytically resulting in a weight update at stage k according to

$$\omega_{k,i} = c_k \cdot \omega_{k-1,i} \cdot \mathcal{N}\left(\hat{\underline{z}}_k; \underline{\mu}^z_{k,i}, \mathbf{C}^z_{k,i}\right)$$

for object class $i = 1 \ldots l$, where $c_k = \left(\sum_i \omega_{k-1,i} \cdot \mathcal{N}\left(\hat{\underline{z}}_k; \underline{\mu}^z_{k,i}, \mathbf{C}^z_{k,i}\right)\right)^{-1}$ is a normalization constant and $\omega_{k-1,i}$ are the weights of $f^e_{k-1}(x)$.

5.3.4 Planning

So far, the camera parameter $\underline{a}_k$ was assumed to be given. But in active object recognition, an action is chosen automatically by the imaging system itself for acquiring high informative observations. For this purpose, the optimization problem

$$\underline{a}^*_k = \arg\max_{\underline{a}_k} \mathrm{I}\left(\boldsymbol{x}, \underline{\boldsymbol{z}}_k | \underline{a}_k\right) \tag{5.26}$$

is formulated to determine the optimal action $\underline{a}^*_k$ to be applied at stage k. Since solving (5.26) results in the camera parameters to be applied next, it is often referred to as *next-best-view* planning (see e.g. [154]). As objective function in (5.26), the *mutual information*

$$\mathrm{I}\left(\boldsymbol{x}, \underline{\boldsymbol{z}}_k | \underline{a}_k\right) = \iint f\left(x, \underline{z}_k\right) \cdot \log \frac{f\left(x, \underline{z}_k\right)}{f(x) \cdot f\left(\underline{z}_k\right)} \,\mathrm{d}x \,\mathrm{d}\underline{z}_k \tag{5.27}$$

between state and feature vector given a camera parameter is considered. This measure quantifies the amount of information the knowledge of an observation reveals about the state and vice versa. It is closely related to Shannon's entropy and zero only iff both variables are independent [46].

Unfortunately, an analytical calculation of the mutual information is not possible as it requires evaluating the logarithm of a Gaussian mixture

Figure 5.9: Cups with different labels.

representing $\underline{z}_k$. To obtain a computationally cheap and robust approximation, the alternative formulation of the mutual information according to

$$\mathrm{I}\left(\boldsymbol{x}, \underline{\boldsymbol{z}}_k | \underline{a}_k\right) = \mathrm{H}\left(\underline{\boldsymbol{z}}_k | \underline{a}_k\right) - \mathrm{H}\left(\underline{\boldsymbol{z}}_k | \boldsymbol{x}, \underline{a}_k\right) \tag{5.28}$$

is employed, where $\mathrm{H}(\boldsymbol{x})$ is the differential entropy

$$\mathrm{H}(\boldsymbol{x}) = -\int f(x) \cdot \log f(x) \,\mathrm{d}x \,.$$

The second term in (5.28) has an analytical expression, while the first term can be bounded from below as shown in [86].

The optimization problem in (5.26) neither is convex nor possesses it a closed-form solution. To increase the probability of finding the optimal camera parameter or at least to ensure finding a parameter that is very close to the optimal one, so-called *multi-start optimization* is performed (see e.g. [180]). Here, optimization is repeated from varying initial points. To cover the camera parameter space $\mathcal{A}$ uniformly, the initial points form a regular grid on $\mathcal{A}$.

Example 29: Recognizing Cups

The object to be recognized in this example are synthetically generated cups as depicted in Figure 5.9. Eight different cups exist, all being identical except for a label that is cut through the surface. The labels of six cups are visible from the same perspective, one is visible from the opposite point of view and one cup is not labeled at all.

For learning and recognition, 100×100 pixel normalized grayscale images are generated from the cups, where zero-mean Gaussian noise with variance 14.7 is added. 1D and 2D features are extracted from the images. In the 1D case, the mean gray value is considered. The eigenspace or principal component decomposition approach proposed in [127] is used for extracting 2D features, where the two largest eigenvalues are taken into account.

By means of the camera parameter, the position can be changed in one or two dimensions. In the 1D case, the camera moves on a circle that is parallel to the horizontal plane and centered at the object. In the 2D case, the camera position can be varied on a sphere centered at the object. Here, the actions correspond to the azimuth and elevation angles.

To learn the GPs, each dimension of the action space is sampled regularly in 10 decimal degree steps, i.e., for the one-dimensional circular action space, this leads to 36 sample images.

For comparison, the following active object recognition approaches are considered:

Planner The proposed approach, where 5 and 15 initial points for optimization are exploited for the 1D and 2D action space, respectively.

Grid An approach similar to [56], where at each stage the action maximizing the mutual information is taken from a finite set. Here, this finite set coincides with the set of initial points of the Planner.

Random Actions are selected uniformly at random.

For each combination of feature and action space, 50 MC simulation runs are performed, where the true cup is selected uniformly at

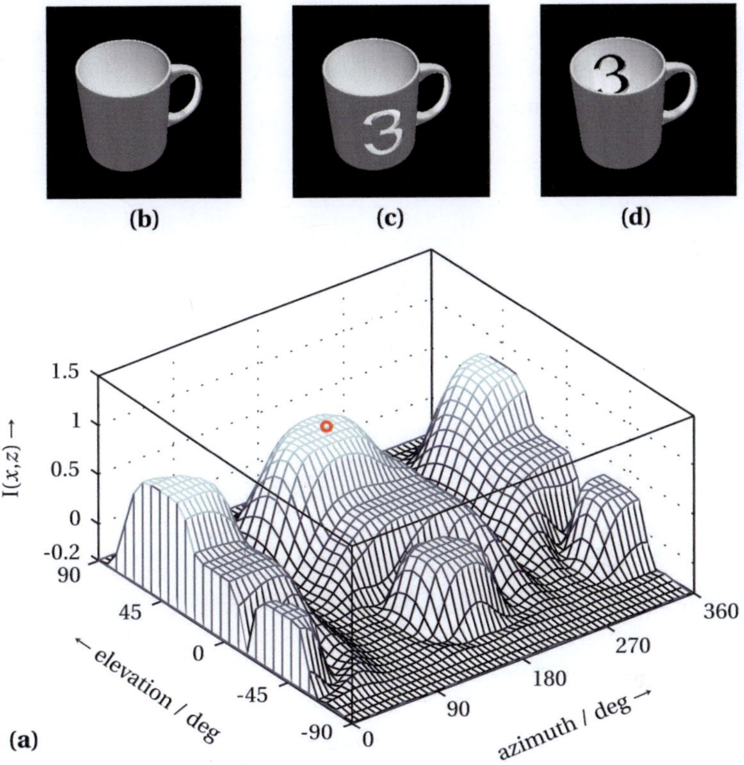

Figure 5.10: (a) Lower bound of mutual information with optimal view/action (red circle). (b)–(d) View of three of the cups corresponding to the optimal action.

random. The initial distribution $f_0^e(x)$ is uniform. A decision about the object type is made if either the probability (weight) of one object class exceeds 0.95 or after eight stages.

For the 2D action space, the mutual information surface for three cups is plotted in Figure 5.10a. Here, the optimal action is indicated by the red circle, which corresponds to an elevation angle of approximately 45°. For this action, the corresponding views on the three

Table 5.1: Cup recognition. (a) recognition rate in percent, (b) average number of views, (c) average maximum object probability.

Dim.	Planner			Grid			Random		
$\mathcal{A}$ / $\mathcal{Z}$	(a)	(b)	(c)	(a)	(b)	(c)	(a)	(b)	(c)
1 / 1	**66**	**6.06**	**0.74**	62	6.1	0.71	50	7.32	0.53
1 / 2	88	**3.08**	**0.97**	74	4.96	0.89	**94**	6.88	0.81
2 / 1	**92**	**2.5**	**0.99**	62	4.1	0.95	76	6.34	0.70
2 / 2	**100**	**1.88**	**0.99**	88	2.5	0.97	68	6.92	0.74

cups are depicted in Figure 5.10b–d. It can be seen that this view facilitates to look inside the cups and thus, allows an easy discrimination of all three cups.

The average values over the 50 simulation runs in terms of recognition rate, number of views, and maximum object probability are listed in Table 5.1. It can be seen that the Planner performs best with respect to almost any performance indicators. In comparison to Random, the number of stages after which a recognition decision is made is significantly lower. Simultaneously, the certainty in this decision is much higher as the average maximum object probability indicates. The performance of the Grid approach is often close to the proposed approach. But the significantly lower number of views of the Planner shows the benefits of performing a continuous optimization for next-best-view planning. In contrast to both Grid and Random, the proposed Planner can take advantage of an increasing feature and action dimension, i.e., with an increasing dimension the recognition rate increases as well and the number of views decreases.

A high object probability not necessarily coincides with the best recognition rate as seen in the case of the 1D action space and 2D feature space. While Random merely relies on the GP object models for estimation, Grid and Planner additionally use the models for planning. Thus, a bootstrapping effect can cause the decision maker to get stuck in a repetitive pattern. The quality of the GP models is essential for the recognition process and thus, under- and over-fitting require special attention.

5.4 Summary

For the applications considered in this chapter, not only the novel algorithms proposed in the previous chapters of this thesis have been turned into practice. Also several additional contributions have been suggested:

- *(Semi-)Analytical measurement updates for position estimation and pose estimation:* By considering a squared range measurement equation a novel computationally efficient closed-form position estimation is derived. In case of pose estimation, this analytical solution can be exploited thanks to the nonlinear-nonlinear decomposition proposed in Section 2.5.2.
- *On-line source estimation based on Gaussian plume model:* The state-of-the-art focuses on off-line MCMC methods for source estimation. By employing the Gaussian plume dispersion model as a measurement model and by modeling the unknown source parameters as state vector with Gaussian mixture density representation facilitates a computationally light-weight but highly accurate source estimation.
- *Gaussian process object models for active object recognition:* Instead of a discretization of the camera parameter space, which is common in the state-of-the-art, the mapping from the latent object class to the feature vector is learned by means of GP regression. This approach can be applied in various recognition scenarios as it is not restricted to specific features, camera parameters, or covariance functions.

- *Efficient next-best-view planning based on lower bound approximation of mutual information:* The GP object models together with a novel lower bound on the mutual information allow optimizing the camera parameters on a fine-grained level and with low computational overhead.

6

Concluding Remarks

6.1 Conclusions

Gaussian filtering lays the foundation to all contributions made in this thesis. In many real-world applications, the Gaussian assumption is valid and thus, sufficient for accurate filtering. In addition, thanks to its algebraical simplicity, Gaussian filtering can be performed in an efficient and scalable manner. The contributions in this thesis exploit closed-form solutions that exist for particular nonlinear models, which leads to a further improvement of the estimation performance and computational efficiency. By means of the proposed decomposition techniques and polynomial approximation, these benefits can even be utilized for problems that are not fully covered by any of the closed-form cases. The experiments and simulations performed for position and pose estimation provide evidence that the proposed Gaussian filtering techniques lead to an improvement compared to state-of-the-art approaches like the EKF or the UKF.

For more complex filtering problems, where the Gaussian assumption no longer holds, the contributions made for Gaussian filtering are not necessarily inapplicable. Quite the contrary, embedding Gaussian filtering into a Gaussian mixture framework turns out to be a very powerful estimation tool. Most of the simplicity of Gaussian filtering remains by this migration. Merely some additional "management" tasks have to be performed for Gaussian mixture filtering. The contributions in this thesis serve these tasks, which are mainly concerned with controlling the number of mixture components. Especially for the proposed adaptive splitting scheme there is experimental evidence that by exploiting the linearization error for introducing components leads to a significant improvement of the estimation performance, while the number of additional components can be kept on a low level. It has been shown that continually adding and removing components is better than filtering with a constantly high number of mixture components. Furthermore, all computations can be performed on-line, which is beneficial over other accurate estimation techniques like MCMC, which only allows off-line or batch processing.

In case of missing mathematical models that describe the system dynamics and sensor characteristics, Gaussian process regression is suggested to learn probabilistic models from data. Given such a GP model, filtering and smoothing can be performed in closed form when restricting to Gaussian distributions. By means of simulations it has been shown that the obtained estimation performance is superior compared to state-of-the-art Gaussian filters, even in cases where these filters utilize the exact model. To bound the computational complexity with a growing data set, a recursive GP regression algorithm has been proposed in addition. Here, the hyperparameters of the GP are learned by means of utilizing the proposed Gaussian filtering techniques, which facilitates on-line learning.

Besides aiming for computationally efficient Bayesian filtering, the contributions made in this thesis lead to a lower user involvement. That is, many of the proposed algorithms can be operated in a black-box fashion. Examples are the automatic Chebyshev series expansion by means of the discrete cosine transform (Section 2.5.3), adaptive Gaussian mixture splitting (Section 3.4.1), automatic model selection in the mixture reduction (Section 3.4.3), or the on-line hyperparameter optimization for GPs (Section 4.5.4). In doing so, the application of these algorithms to a given problem can be simplified, which reduces deployment time and operational costs.

6.2 Future Work

In all three pillars considered in this thesis there is enough room for further improvements and extensions. In the following, an outlook on future work is provided.

Gaussian Filtering

One of the contributions in this thesis that currently requires a significant amount of manual inspection is the nonlinear-nonlinear decomposition in Section 2.5.2. An expert has to investigate manually which parts of the state are analytically integrable and which are not. To facilitate an automatic decomposition, one idea is to utilize Risch's algorithm[1] [150, 151] together with a decomposition exploration algorithm similar to the one proposed in [103].

The moment homotopy for polynomial nonlinearities in Section 2.5.5 is a first step towards removing the joint Gaussian assumption between state and measurement. In general, it is desirable to perform Gaussian filtering without this assumption for arbitrary nonlinearities to improve the robustness and to reduce the estimation error. First approaches in this direction can be found in [75, 106].

The measurement update of the Gaussian filter for polynomial nonlinearities can naturally output a full exponential density representation.

1 More precisely, a realization of this algorithm in modern computer algebra systems.

Maintaining this representation also over the prediction, however, is more difficult as neither the predicted density nor even the predicted moments can be expressed analytically. Even if the predicted moments were available, determining an exponential density that matches the moments is also not possible in closed form. See [148] for a first step to solve this issue.

The CPKF proposed in Section 2.5.3 is operational for one-dimensional states so far. The next step is to extend it to multiple dimensions. While this is straightforward in terms of the Chebyshev series expansion, closed-form moment calculation for Gaussians mapped through multi-dimensional polynomials is still computationally demanding. The results in [98] already lowered the computations significantly compared to previous approaches, but the moment recursion proposed in Section 2.5.4 might offer a way to a further reduction.

Gaussian Mixture Filtering

Thanks to the individual processing of the Gaussian components considered in this thesis, Gaussian mixture filters directly benefit from any improvement achieved for Gaussian filters. Thus, the outlook on future work for mixture filters is mainly focused on the refinement and reapproximation operations of Algorithm 3.

The SGMR algorithm proposed in Section 3.4.3 so far is only applicable for one-dimensional and two-dimensional mixtures due to the use of the curvature as roughness penalty. Thus, future work is dedicated to explore and propose curvature measures for higher dimensions.

Splitting a Gaussian component into many as discussed in Section 3.4.2 takes the linearization error into account. This procedure is generally applicable, but can be enhanced by application-specific criteria. For instance, for the source estimation application considered in Section 5.2, it might be beneficial to also take the distance between sensor location and component mean into account. This avoids situations where no component is split due to large distances—the Gaussian plume model is then approximately linear. In this case, the AGMF degenerates to a simple Gaussian mixture filter with a fixed number of components.

The source estimation application gives room for further improvements. Currently it is assumed that the number of sources is known a priori, but actually new sources might appear spontaneously or an existing source might disappear over time. Such birth and death processes being common in multiple target tracking (see e.g. [186]) should be incorporated.

Gaussian Process Filtering

For the GP-ADF and GP-RTSS proposed in Section 4.5.1 and Section 4.5.2, respectively, it is assumed that training data of the state is given, but as this state is hidden, this assumption is impractical for many applications. The GP for the system and the measurement model has to be learned without the need of direct access to the hidden states. This can be achieved by means of Expectation Maximization since both GP-ADF and GP-RTSS allow for gradient-based parameter optimization.

The number and placement of the basis vectors required for the RGP in Section 4.5.3 has not been discussed. The algorithm supports adding and removing basis vectors on-line, which allows correcting an insufficient initial selection of basis vectors. However, a criterion that facilitates a good choice of new basis vectors is left for future work. Techniques used in active learning [110] or sensor planning [83] for instance can be utilized for this purpose.

Rather straightforward to extend is the on-line hyperparameter learning proposed in Section 4.5.4. Instead of restricting to a Gaussian representation of the hyperparameters, also Gaussian mixtures can be used by means of exploiting the techniques proposed in Chapter 3.

A

Particle Filtering

Monte Carlo (MC) methods for solving the integrals appearing in Bayesian filtering became popular from the 1980s on, with the advent of cheap but powerful micro-processors. In contrast to the previously popular Kalman filtering methods and its derivatives, MC methods make no assumptions regarding the models or density functions. In the following, a brief introduction to particle filters is given, which are a popular form of MC approximations to Bayesian filtering.

A.1 Perfect Monte Carlo Sampling

MC methods rely on a non-parametric representation of the density function $f(\underline{x} \,|\, \hat{\underline{z}}_{0:k})$ by means of a set of n independent and identically distributed samples $\underline{x}^{(i)}$, $i = 1 \ldots n$. These samples are often named particles, which led to the naming *particle filters*. Given the sample set, the density function can be approximated as a sum of Dirac delta distributions according to

$$f(\underline{x}\,|\,\hat{\underline{z}}_{0:k}) \approx \tfrac{1}{n}\sum_{i=1}^{n} \delta(\underline{x}-\underline{x}_i)\,. \tag{A.1}$$

For an arbitrary nonlinear function $\underline{g}(\underline{x}) : \mathbb{R}^{n_x} \to \mathbb{R}^{n_y}$, this representation leads to a perfect MC approximation of the expectation calculation

$$\mathrm{E}\{\underline{g}(\underline{\boldsymbol{x}})\} = \int \underline{g}(\underline{x}) \cdot f(\underline{x}\,|\,\hat{\underline{z}}_{0:k})\,\mathrm{d}\underline{x} \overset{(A.1)}{\approx} \tfrac{1}{n}\sum_{i=1}^{n} \underline{g}(\underline{x}^{(i)})\,. \tag{A.2}$$

The central limit theorem guarantees that the MC approximation converges with an increasing number of particles, regardless of the dimension of $\underline{\boldsymbol{x}}$. This dimensionless property is unique to MC methods compared to other (deterministic) numerical integration methods, at least from a theoretical point of view [59]. However, practice shows that the number of required particles also grows exponentially with the dimension of $\underline{\boldsymbol{x}}$ (see e.g. [49]).

Despite the nice theoretical properties of MC approximation, sampling from $f(\underline{x}\,|\,\hat{\underline{z}}_{0:k})$ is often very difficult as the density typically has a complicated functional form and is only known up to a normalization constant. A solution to this issue is *importance sampling.*

A.2 Importance Sampling

The key idea of importance sampling is to use an approximation density $\pi(\underline{x}\,|\,\hat{\underline{z}}_{0:k})$ called *importance function* instead of $f(\underline{x}\,|\,\underline{z}_{1:k})$. Samples can be drawn much easier from the importance function [184]. If it holds that $\pi(\underline{x}\,|\,\hat{\underline{z}}_{0:k}) > 0$ whenever $f(\underline{x}\,|\,\hat{\underline{z}}_{0:k}) > 0$ then the expectation in (A.2) can be decomposed to

$$\mathrm{E}\{\underline{g}(\underline{\boldsymbol{x}})\} = \int \left(\underline{g}(\underline{x})\frac{f(\underline{x}\,|\,\hat{\underline{z}}_{0:k})}{\pi(\underline{x}\,|\,\hat{\underline{z}}_{0:k})}\right)\pi(\underline{x}\,|\,\hat{\underline{z}}_{0:k})\,\mathrm{d}\underline{x} = \mathrm{E}\{\underline{g}(\underline{\boldsymbol{x}})\cdot\omega(\underline{\boldsymbol{x}})\}\,,$$

with weight $\omega(\underline{x}) \triangleq \frac{f(\underline{x}|\hat{\underline{z}}_{0:k})}{\pi(\underline{x}|\hat{\underline{z}}_{0:k})}$. By now drawing samples $\underline{x}_i$ from the importance function and not from the density $f(\underline{x}|\hat{\underline{z}}_{0:k})$, the expectation in (A.2) can be approximated as

$$\mathrm{E}\left\{\underline{g}(\boldsymbol{\underline{x}})\right\} = \sum_{i=1}^{n} \omega^{(i)} \cdot \underline{g}(\underline{x}^{(i)})$$

with normalized weights

$$\omega^{(i)} = \frac{\omega(\underline{x}^{(i)})}{\sum_{i=1}^{n} \omega(\underline{x}^{(i)})} \tag{A.3}$$

Thus, the set of particles now comprises the particles itself and the corresponding weights (A.3). Accordingly, the density is approximated by means of a weighted sum of Dirac delta distributions

$$f(\underline{x}|\hat{\underline{z}}_{0:k}) \approx \sum_{i=1}^{n} \omega^{(i)} \cdot \delta(\underline{x} - \underline{x}^{(i)}) .$$

A.2.1 Sequential Importance Sampling

Importance sampling as introduced above does not allow for recursive filtering as required for models of the form

$$\begin{aligned} \boldsymbol{\underline{x}}_k &\sim f(\underline{x}_k|\underline{x}_{k-1}) , \\ \boldsymbol{\underline{z}}_k &\sim f(\underline{z}_k|\underline{x}_k) , \end{aligned}$$

where the state $\boldsymbol{\underline{x}}_k$ varies with the time k. Instead, one would have to recalculate the weights whenever new measurements become available, which leads to a growing computational demand with an increasing number of measurements and time, respectively. The main reason of this drawback lays in the definition of the importance function. For recursive processing, the importance function itself has to follow a recursion according to

$$\pi\left(\underline{x}_{0:k} | \hat{\underline{z}}_{0:k}\right) = \pi\left(\underline{x}_k | \underline{x}_{0:k-1}, \hat{\underline{z}}_{0:k}\right) \cdot \pi\left(\underline{x}_{0:k-1} | \hat{\underline{z}}_{0:k-1}\right) ,$$

which leads to a recursive expression for the weights

$$\omega_k^{(i)} \propto \frac{f\left(\hat{\underline{z}}_k | \underline{x}_k^{(i)}\right) \cdot f\left(\underline{x}_k^{(i)} | \underline{x}_{k-1}^{(i)}\right)}{\pi\left(\underline{x}_k^{(i)} | \underline{x}_{0:k-1}^{(i)}, \hat{\underline{z}}_{0:k}\right)} \cdot \underbrace{\frac{f\left(\underline{x}_{0:k-1} | \hat{\underline{z}}_{0:k-1}\right)}{\pi\left(\underline{x}_{0:k-1}^{(i)} | \hat{\underline{z}}_{0:k-1}\right)}}_{\propto\, \omega_{k-1}^{(i)}} . \tag{A.4}$$

This recursion follows from the observation that the samples $\underline{x}_{0:k-1}^{(i)}$ have already been drawn from the importance function $\pi\left(\underline{x}_{0:k-1} | \hat{\underline{z}}_{0:k-1}\right)$ and the weights $\omega_{k-1}^{(i)}$ have already been calculated in the time step $k-1$. Thus, the samples $\underline{x}_{0:k}^{(i)}$ can been obtained from $\pi\left(\underline{x}_{0:k} | \hat{\underline{z}}_{0:k}\right)$ by drawing $\underline{x}_k^{(i)}$ from $\pi\left(\underline{x}_k | \underline{x}_{0:k-1}^{(i)}, \hat{\underline{z}}_{0:k}\right)$. This efficient MC approximation of drawing samples and calculating the weights according to (A.4) is called *sequential importance sampling* (SIS), which leads to the approximation

$$f_k^e\left(\underline{x}_k\right) \approx \sum_{i=1}^{n} \omega_k^{(i)} \cdot \delta\left(\underline{x}_k - \underline{x}_k^{(i)}\right) \tag{A.5}$$

of the posterior density of the state $\underline{\boldsymbol{x}}_k$.

A.2.2 Choice of Importance Function

The SIS is still formulated quite generally as the concrete choice of the importance function leaves many degrees of freedom. A common further very practical restriction made is to apply the Markov assumption, which leads to

$$\pi\left(\underline{x}_k | \underline{x}_{0:k-1}, \hat{\underline{z}}_{0:k}\right) = \pi\left(\underline{x}_k | \underline{x}_{k-1}, \hat{\underline{z}}_{0:k}\right) .$$

In doing so, it is no longer necessary to store the whole particle trajectory $\underline{x}_{0:k}^{(i)}$, but only the current particles $\underline{x}_k^{(i)}$. It can be shown, that the *optimal importance function* minimizing the variance of the weights $\omega_k^{(i)}$ is given by

$$\pi\left(\underline{x}_k | \underline{x}_{k-1}, \hat{\underline{z}}_{0:k}\right) = f\left(\underline{x}_k | \underline{x}_{k-1}, \hat{\underline{z}}_k\right) ,$$

but unfortunately the optimal importance function is typically not given in analytic form and drawing samples from it is not possible [60]. One approximation often applied to circumvent this issue is it utilize Gaussian filtering techniques like linearization (see Section 2.2.3) or linear regression (see Section 2.2.5), which leads for instance to the extended particle filter or unscented particle filter [201].

Another variation of SIS is the employ the transition density $f\left(\underline{x}_k | \underline{x}_{k-1}\right)$ as importance function. This choice leads to the *bootstrap filter* [71] allowing a very simple implementation as drawing the particles $\underline{x}_k^{(i)}$ corresponds to evaluating the system function $\underline{a}_k(.)$ on the given particles $\underline{x}_{k-1}^{(i)}$ and samples from the system noise $\underline{\boldsymbol{w}}_k$. The drawback of the bootstrap filter is that no measurement information is used for drawing $\underline{x}_k^{(i)}$, which typically leads to a very large number of particles for accurate estimates.

A.2.3 Resampling

By applying SIS it easily happens that over time almost all particles have a weight of (nearly) zero, which effectively reduces the number of particles. To overcome this *sample degeneration* problem a so-called *resampling* step has to be applied in addition. Here, after performing the weight update according to (A.4), n new particles are drawn from the posterior (A.5) that replace the current particle set. The new particles all have the same weight $1/n$.

Over the years many resampling methods have been proposed (see e.g. [37, 102, 115]). The key idea of any resampling is to remove particles with small weights and duplicate those with a large weight. MC methods with

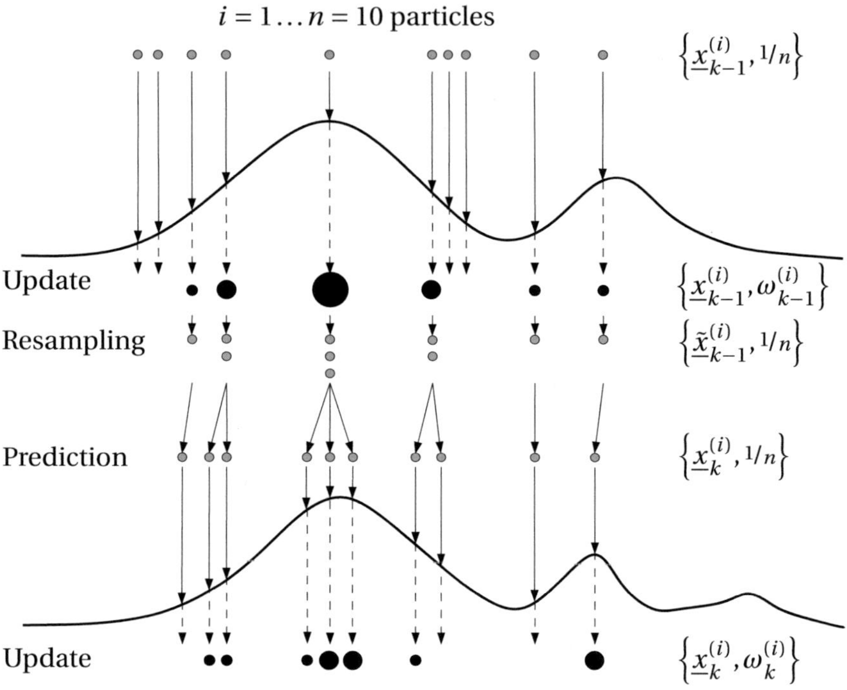

Figure A.1: SIR starts with a particle set $\left\{\underline{x}_{k-1}^{(i)}, 1/n\right\}$ at time step $k-1$. The weight of each particle is updated given the current measurement value $\underline{\hat{z}}_{k-1}$, which results in the particle set $\left\{\underline{x}_{k-1}^{(i)}, \omega_{k-1}^{(i)}\right\}$ representing the posterior $f_{k-1}^{e}(\underline{x}_k)$. The resampling step duplicates the particles with high weights. The resulting particle set $\left\{\underline{\tilde{x}}_{k-1}^{(i)}, 1/n\right\}$ still represents the posterior $f_{k-1}^{e}(\underline{x}_k)$. Drawing the particles $\underline{x}_k^{(i)}$ from the importance function, which corresponds to the prediction step, leads to the particle set $\left\{\underline{x}_{k}^{(i)}, 1/n\right\}$ representing the predicted density $f_k^p(\underline{x}_k)$. (Image adapted from [59])

an additional resampling step are called *sequential importance resampling* (SIR), which are more known under the term *particle filter* (PF). In Figure A.1 the steps of a PF are depicted.

B

Performance Measures

In this chapter, some measures quantifying the estimation performance of Bayesian filtering algorithms are briefly introduced.

B.1 Root Mean Square Error

Generally, the error between the estimated mean $\underline{\mu}_k^x$ and the true state $\underline{x}_k$ is defined as

$$\underline{e}_k \triangleq \underline{x}_k - \underline{\mu}_k^x , \tag{B.1}$$

which is a vector of dimension-wise errors $e_{k,i} = x_{k,i} - \mu_{k,i}^x$ for $i = 1 \ldots n_x$. A standard performance measure for Bayesian filters depending on the error (B.1) is the *root mean square error* (rmse) defined by

$$\text{rmse}_i \triangleq \sqrt{\frac{1}{K} \sum_{k=1}^{K} e_{k,i}^2}$$

for dimension $i = 1 \dots n_x$, where K is the number of time steps.

In this thesis, MC simulations are used besides real data in order to evaluate the performance of a filter. In case of MC simulations, the rmse for each time step over the different simulations can be evaluated. Hence, the rmse for time step k and dimension i is given by

$$\mathrm{rmse}_i^{\mathrm{MC}} \triangleq \sqrt{\frac{1}{n_{\mathrm{MC}}} \sum_{m=1}^{n_{\mathrm{MC}}} \left(e_{k,i}^m\right)^2}$$

where $e_{k,i}^m$ is the estimation error of dimension i for the mth MC simulation run and n_{MC} is the number of MC simulation runs.

B.2 Mean Absolute Error

A performance measure very similar to the rmse is the *mean absolute error* (mae), which is defined as

$$\mathrm{mae}_i \triangleq \frac{1}{K} \sum_{k=1}^{K} \left|e_{k,i}\right| \tag{B.2}$$

for dimension $i = 1 \dots n_x$.

B.3 Normalized Estimation Error Square

The rmse and mae merely takes the mean of the estimated state into account. The *normalized estimation error square* (nees) considers the uncertainty of the estimation in addition. It is defined as

$$\mathrm{nees}_k \triangleq \left(\underline{x}_k - \underline{\mu}_k^x\right)^{\mathrm{T}} \left(\mathbf{C}_k^x\right)^{-1} \left(\underline{x}_k - \underline{\mu}_k^x\right) ,$$

which corresponds to a weighted Euclidean distance of the state errors at time step k, where the weight is given by the inverse state covariance matrix. Due to incorporating the inverse covariance matrix, a large estimation error has a small contribution to the nees if the covariance is large and conversely, a large estimation error contributes more in case of a small covariance. Thus, this measure allows indicating a consistent filter.

In contrast to the rmse, the nees is a dimensionless quantity. It is χ^2-distributed with n_x degrees of freedom if the state estimate is Gaussian [72]. The nees is also known as *Mahalanobis distance.*

B.4 Negative Log-Likelihood

An alternative to the nees that is often used in machine learning is the *negative log-likelihood* (nll)

$$\begin{aligned} &\text{(filtering)} \quad \mathrm{nll}_k \triangleq -\log f_k^x(\underline{x}_k)\,, \\ &\text{(learning)} \quad \mathrm{nll} \triangleq -\sum_{i=1}^{n} \log f(\underline{\hat{z}}_i | \underline{x})\,, \end{aligned} \tag{B.3}$$

where the first term is used for performance measurement in filtering, while the second term is used in a model learning context e.g. via GPs in this thesis. The dependence of the latter definition on the likelihood $f(\underline{\hat{z}}|\underline{x})$ explains the naming of this performance measure. In general, the nll quantifies how well a realization (true state or measurement value) can be explained by the given (estimated or learned) density function.

In case of a Gaussian state density, the first nll term in (B.3) can be formulated to

$$\mathrm{nll}_k = \log\sqrt{|2\pi \mathbf{C}_k^x|} + \tfrac{1}{2}\left(\underline{x}_k - \underline{\mu}_k^x\right)^{\mathrm{T}} \left(\mathbf{C}_k^x\right)^{-1} \left(\underline{x}_k - \underline{\mu}_k^x\right) .$$

The second term in (B.3) can be resolved similarly. It can be see that the nll consists of the nees and a term that penalizes a too large covariance matrix, which allows discovering of overestimation.

C

Quadratic Programming

Optimization problems with quadratic objective function and affine constraints are called *quadratic programs* (QPs). Hence, a quadratic program is defined as

$$\begin{aligned} \min_{\underline{x}} \quad & \tfrac{1}{2} \cdot \underline{x}^{\mathrm{T}} \mathbf{Q} \underline{x} + \underline{q}^{\mathrm{T}} \cdot \underline{x} + c \\ \text{s.t.} \quad & \mathbf{G} \cdot \underline{x} \preceq \underline{h} \\ & \mathbf{A} \cdot \underline{x} = \underline{b} \, , \end{aligned} \tag{C.1}$$

where $\mathbf{Q}$ is a symmetric matrix and c is a constant. Solving a QP generally is NP-hard [157]. However, if the matrix $\mathbf{Q}$ is in addition positive definite, a QP becomes a special case of a convex optimization problem and thus, any locally optimal solution to (C.1) is also globally optimal. This optimal solution can be determined in polynomial time for instance by means of an ellipsoid method. For further reading on QP and convex optimization in general see [24].

Bibliography

[1] Fahed Abdallah, Amadou Gning, and Philippe Bonnifait. Box Particle Filtering for Nonlinear State Estimation using Interval Analysis. *Automatica*, 44(3):807–815, March 2008.

[2] Hervé Abdi and Lynne J. Williams. Principal Component Analysis. In *Wiley Interdisciplinary Reviews: Computational Statistics*, volume 2, pages 433–459. Wiley, New York, July 2010.

[3] Simo Ali-Löytty. Efficient Gaussian Mixture Filter for Hybrid Positioning. In *Proceedings of the IEEE/ION Position, Location and Navigation Symposium*, pages 60–66, Monterey, CA, May 2008.

[4] Simo Ali-Löytty. *Gaussian Mixture Filters in Hybrid Positioning.* PhD thesis, Tampere University of Technology, Tampere, Finland, August 2009.

[5] Simo Ali-Löytty and Niilo Sirola. Gaussian Mixture Filter in Hybrid Navigation. In *Proceedings of the European Navigation Conference*, pages 831–837, May 2007.

[6] Simo Ali-Löytty and Niilo Sirola. Gaussian Mixture Filters for Hybrid Positioning. In *Proceedings of the 20th International Technical Meeting of the Satellite Devision of the Institute of Navigation (ION GNSS)*, pages 562–569, Fort Worth, TX, September 2007.

[7] Daniel L. Alspach and Harold W. Sorenson. Nonlinear Bayesian Estimation using Gaussian Sum Approximation. *IEEE Transactions on Automatic Control*, 17(4):439–448, August 1972.

[8] Brian D. O. Anderson and John B. Moore. *Optimal Filtering*. Dover Publications, 2005.

[9] Christophe Andrieu and Arnaud Doucet. Particle filtering for partially observed Gaussian state space models. *Journal of the Royal Statistical Society: Series B*, 64(4):827–836, 2002.

[10] Ienkaran Arasaratnam. *Cubature Kalman Filtering: Theory & Applications*. PhD thesis, McMaster University, April 2009.

[11] Ienkaran Arasaratnam and Simon Haykin. Cubature Kalman Filters. *IEEE Transactions on Automatic Control*, 54(6):1254–1269, June 2009.

[12] Ienkaran Arasaratnam and Simon Haykin. Cubature Kalman Smoothers. *Automatica*, 47(20):2245–2250, October 2011.

[13] Ienkaran Arasaratnam, Simon Haykin, and Robert J. Elliott. Discrete-Time Nonlinear Filtering Algorithms Using Gauss–Hermite Quadrature. *Proceedings of the IEEE*, 95(5):953–977, 2007.

[14] M. Sanjeev Arulampalam, Simon Maskell, Neil Gordon, and Tim Clapp. A Tutorial on Particle Filters for Online Nonlinear/Non-Gaussian Bayesian Tracking. *IEEE Transactions on Signal Processing*, 50(2):174–188, February 2002.

[15] S. Bancroft. An Algebraic Solution of the GPS Equations. *IEEE Transactions on Aerospace and Electronic Systems*, AES-21(1):56–59, January 1985.

[16] Richard E. Bellman. *Dynamic Programming*. Princeton University Press, 1957.

[17] Frederik Beutler, Marco F. Huber, and Uwe D. Hanebeck. Instantaneous Pose Estimation using Rotation Vectors. In *Proceedings of the 34th International Conference on Acoustics, Speech, and Signal Processing (ICASSP)*, pages 3413–3416, Taipei, Taiwan, April 2009.

[18] Frederik Beutler, Marco F. Huber, and Uwe D. Hanebeck. Semi-Analytic Stochastic Linearization for Range-Based Pose Tracking. In *Proceedings of the 2010 IEEE International Conference on Multisensor Fusion and Integration for Intelligent Systems (MFI)*, pages 44–49, Salt Lake City, Utah, September 2010.

[19] Jürgen Beyerer. *Verfahren zur quantitativen statistischen Bewertung von Zusatzwissen in der Messtechnik.* VDI Fortschritt-Berichte, Reihe 8, Nummer 783, 1999.

[20] Samuel S. Blackman. *Multiple-Target Tracking with Radar Applications.* Norwood, MA: Artech House, 1986.

[21] Hermann Borotschnig, Lucas Paletta, Manfred Prantl, and Axel Pinz. Appearance-Based Active Object Recognition. *Image and Vision Computing*, 18:715–727, 2000.

[22] John E. Bortz. A New Mathematical Formulation for Strapdown Inertial Navigation. *IEEE Transactions on Aerospace and Electronic Systems*, AES-7(1):61–66, January 1971.

[23] Mieczyslaw Borysiewicz, Anna Wawrzynczak, and Piotr Kopka. Bayesian-Based Methods for the Estimation of the Unknown Model's Parameters in the Case of the Localization of the Atmospheric Contamination Source. *Foundations of Computing and Decision Sciences*, 37(4):253–270, 2012.

[24] Stephen Boyd and Lieven Vandenberghe. *Convex Optimization.* Cambridge University Press, 2004.

[25] Philipp Boyle and Marcus Frean. Dependent Gaussian Processes. In Lawrence K. Saul, Yair Weiss, and Leon Bottou, editors, *Advances*

in Neural Information Processing Systems 17, pages 217–224. MIT Press, 2005.

[26] Kai P. Briechle. *Nichtlineare Filterverfahren mit Anwendung auf Lokalisierungsprobleme.* PhD thesis, Technische Universität München, March 2003.

[27] Damiano Brigo and Francois Le Gland. A Finite Dimensional Filter with Exponential Density. In *Proceedings of the 1997 IEEE Conference on Decision and Control (CDC)*, volume 2, pages 1643–1644, San Diego, CA, 1997.

[28] Damiano Brigo, Bernard Hanzon, and Francois Le Gland. A Differential Geometric Approach to Nonlinear Filtering: the Projection Filter. Technical Report 2598, Insitut National De Recherche en Informatique et en Automatique, June 1995.

[29] Damiano Brigo, Bernard Hanzon, and Francois Le Gland. Approximate Nonlinear Filtering by Projection on Exponential Manifold of Densities. *Bernoulli*, 5(3):495–534, June 1999.

[30] Vladimir Britanak, Patrick C. Yip, and Kamisetty R. Rao. *Discrete Cosine and Sine Transforms: General Properties, Fast Algorithms and Integer Approximations.* Academic Press, 2006.

[31] Pierrick Bruneau, Marc Gelgon, and Fabien Picarougne. Parsimonious reduction of Gaussian mixture models with a variational-Bayes approach. *Pattern Recognition*, 43:850–858, 2010.

[32] Lucian Buşoniu, Robert Babuska, Bart De Schutter, and Damien Ernst. *Reinforcement Learning and Dynamic Programming Using Function Approximators.* CRC Press, 2010.

[33] Richard S. Bucy and Kenneth D. Senne. Digital synthesis of nonlinear filters. *Automatica*, 7(3):287–298, May 1971.

[34] James J. Caffery, Jr. A New Approach to the Geometry of TOA Location. In *Proceedings of 55th IEEE Vehicular Technology Conference*, pages 1942–1949, 2000.

[35] Yanshuai Cao and David J. Fleet. Generalized Product of Experts for Automatic and Principled Fusion of Gaussian Process Predictions. In *Modern Nonparametrics 3: Automating the Learning Pipeline workshop at NIPS*, Montreal, Quebec, Canada, December 2014.

[36] Manfredo Do Carmo. *Differential Geometry of Curves and Surfaces.* Prentice-Hall, Englewood Cliffs, NJ, 1976.

[37] James Carpenter, Peter Clifford, and Paul Fearnhead. Improved particle filter for nonlinear problems. In *IEE Proceedings Radar, Sonar and Navigation*, volume 146, pages 2–7, February 1999.

[38] M. D. Carrascal, M. Puigcerver, and P. Puig. Sensitivity of Gaussian plume model to dispersion specifications. In *Theoretical and Applied Climatology*, volume 48, pages 147–157. Springer, 1993.

[39] Subhash Challa, Yaakov Bar-Shalom, and Vikram Krishnamurthy. Nonlinear Filtering via Generalized Edgeworth Series and Gauss-Hermite Quadrature. *IEEE Transactions on Signal Processing*, 48(6):1816–1820, June 2000.

[40] K. C. Chang and Wei Sun. Scalable Fusion with Mixture Distributions in Sensor Networks. In *Proceedings of the 11th International Conference Control, Automation, Robotics and Vision*, pages 1252–1256, Singapore, December 2010.

[41] Rong Chen and Jun S. Liu. Mixture Kalman Filters. *Journal of the Royal Statistical Society: Series B*, 62(3):493–508, 2000.

[42] Zhe Chen. Bayesian Filtering: From Kalman Filters to Particle Filters, and Beyond. Technical report, Adaptive Systems Laboratory, McMaster University, 2003.

[43] Kwok-Wai Cheung and Hing Cheung So. A Multidimensional Scaling Framework for Mobile Location Using Time-of-Arrival Measurements. *IEEE Transactions on Signal Processing*, 53(2):460–470, February 2005.

[44] Martin Clark and Richard Vinter. A New Class of Moment Matching Filters for Nonlinear Tracking and Estimation Problems. In *2006 IEEE Nonlinear Statistical Signal Processing Workshop*, pages 108–112, September 2006.

[45] C. W. Clenshaw. A note on the summation of Chebyshev series. *Mathematical Tables and other Aids to Computation*, 9(51):118–120, 1955.

[46] Thomas M. Cover and Joy A. Thomas. *Elements of Information Theory*. John Wiley & Sons, Inc., 1991.

[47] David F. Crouse, Peter Willett, Krishna Pattipati, and Lennart Svensson. A Look At Gaussian Mixture Reduction Algorithms. In *Proceedings of the 14th International Conference on Information Fusion*, Chicago, IL, July 2011.

[48] Lehel Csató and Manfred Opper. Sparse on-line Gaussian processes. *Neural Computation*, 14(3):641–668, March 2002.

[49] Fred Daum and Jim Huang. Curse of Dimensionality and Particle Filters. In *Proceedings of the 2003 IEEE Aerospace Conference*, volume 4, pages 1979–1993, March 2003.

[50] Guido de Croon, Ida G. Sprinkhuizen-Kuyper, and Eric O. Postma. Comparing Active Vision Models. *Image and Vision Computing*, 27:374–384, March 2009.

[51] Frank Deinzer, Joachim Denzler, and Heinrich Niemann. Viewpoint Selection - Planning Optimal Sequences of Views for Object Recognition. In *In International Conference on Computer Vision*, pages 65–73. Springer, 2003.

[52] Marc P. Deisenroth. *Efficient Reinforcement Learning Using Gaussian Processes.* PhD thesis, Karlsruhe Institute of Technology, 2010.

[53] Marc P. Deisenroth, Marco F. Huber, and Uwe D. Hanebeck. Analytic Moment-based Gaussian Process Filtering. In *26th International Conference on Machine Learning (ICML)*, pages 225–232, Montreal, Canada, June 2009.

[54] Marc P. Deisenroth, Ryan Turner, Marco F. Huber, Uwe D. Hanebeck, and Carl E. Rasmussen. Robust Filtering and Smoothing with Gaussian Processes. *IEEE Transactions on Automatic Control*, 57(7):1865–1871, July 2012.

[55] Marc Peter Deisenroth and Henrik Ohlsson. A General Perspective on Gaussian Filtering and Smoothing: Explaining Current and Deriving New Algorithms. In *Proceedings of the American Control Conference 2011*, pages 1807–1812, San Francisco, CA, June 2011.

[56] Joachim Denzler and Christopher M. Brown. Information Theoretic Sensor Data Selection for Active Object Recognition and State Estimation. *IEEE Transactions on Pattern Analysis and Machine Intelligence*, 24(2):145–157, February 2002.

[57] Peter J. Diggle and Paulo J. Ribeiro, Jr. *Model-Based Geostatistics.* Springer Series in Statistics. Springer, 2007.

[58] Pedro Domingos and Michael Pazzani. On the Optimality of the Simple Bayesian Classifier under Zero-One Loss. *Machine Learning*, 29:103–130, 1997.

[59] Arnaud Doucet, Nando de Freitas, and Neil Gordon. *Sequential Monte Carlo Methods in Practice.* Statistics for Engineering and Information Science. New York: Springer-Verlag, 2001.

[60] Arnaud Doucet, Simon J. Godsill, and Christophe Andrieu. On Sequential Monte Carlo Sampling Methods for Bayesian Filtering. *Statistics and Computing*, 10:197–208, 2000.

[61] Richard O. Duda, Peter E. Hart, and David G. Stork. *Pattern Classification.* Wiley & Sons, 2nd edition, 2000.

[62] Yaakov Engel, Shie Mannor, and Ron Meir. The Kernel Recursive Least-Squares Algorithm. *IEEE Transactions on Signal Processing,* 52(8):2275–2285, August 2004.

[63] Friedrich Faubel and Dietrich Klakow. An Adaptive Level of Detail Approach to Nonlinear Estimation. In *Proceedings of the 2010 IEEE International Conference on Acoustics, Speech and Signal Processing (ICASSP),* pages 3958–3961, 2010.

[64] Friedrich Faubel, John McDonough, and Dietrich Klakow. The Split and Merge Unscented Gaussian Mixture Filter. *IEEE Signal Processing Letters,* 16(9):786–789, September 2009.

[65] Brian Ferris, Dirk Hähnel, and Dieter Fox. Gaussian Processes for Signal Strength-Based Location Estimation. In *Proceedings of Robotics Science and Systems,* 2006.

[66] Wade Foy. Position-Location Solutions by Taylor-Series Estimation. *IEEE Transactions on Aerospace and Electronic Systems,* AES-12(2):187–194, March 1976.

[67] Arthur Gelb. *Applied Optimal Estimation.* MIT Press, 1974.

[68] Agathe Girard, Carl E. Rasmussen, Joaquin Quiñonero-Candela, and Roderick Murray-Smith. Gaussian Process PriorsWith Uncertain Inputs Application to Multiple-Step Ahead Time Series Forecasting. In *Advances in Neural Information Processing Systems 15,* 2003.

[69] Agathe Girard, Carl Edward Rasmussen, and Roderick Murray-Smith. Gaussian Process priors with Uncertain Inputs: Multiple-Step-Ahead Prediction. Technical Report TR-2002-119, University of Glasgow, Department of Computing Science, October 2002.

[70] Jacob Goldberger and Sam Roweis. Hierarchical Clustering of a Mixture Model. In *Proceedings of Neural Information Processing Systems (NIPS)*, pages 505–512, 2005.

[71] Neil J. Gordon, David J. Salmond, and Adrian F. M. Smith. Novel approach to nonlinear/non-Gaussian Bayesian state estimation. *IEE Proceedings on Radar and Signal Processing*, 140(2):107–113, April 1993.

[72] Karl Granström. *Extended target tracking using PHD filters*. PhD thesis, Linköping University, 2012.

[73] Alexander G. Gray and Andrew W. Moore. 'N-Body' Problems in Statistical Learning. In *Advances in Neural Information Processing Systems 13*, 2001.

[74] Uwe D. Hanebeck. *Nonlinear Methods for State Estimation in Stochastic Dynamical Systems – A Concise Introduction*. Habilitation treatise, Technische Universität München, 2002.

[75] Uwe D. Hanebeck. PGF 42: Progressive Gaussian Filtering with a Twist. In *Proceedings of the 16th International Conference on Information Fusion*, Istanbul, Turkey, July 2013.

[76] Uwe D. Hanebeck, Kai Briechle, and Andreas Rauh. Progressive Bayes: A New Framework for Nonlinear State Estimation. In *Proceedings of SPIE, AeroSense Symposium*, volume 5099, pages 256–267, Orlando, Florida, May 2003.

[77] Uwe D. Hanebeck and Olga Feiermann. Progressive Bayesian Estimation for Nonlinear Discrete-Time Systems: The Filter Step for Scalar Measurements and Multidimensional States. In *Proceedings of the 2003 IEEE Conference on Decision and Control*, pages 5366–5371, Maui, Hawaii, USA, December 2003.

[78] Uwe D. Hanebeck and Günther Schmidt. Closed-Form Elliptic Location with an Arbitrary Array Topology. In *Proceedings of IEEE*

International Conference on Acoustics, Speech, and Signal Processing (ICASSP), pages 3070–3073, 1996.

[79] Steven Hanna, Joseph Chang, and Helge R. Olesen. *Indianapolis Tracer Data and Meteorological Data*, May 1997.

[80] Bill Hirst, Philip Jonathan, Fernando G. del Cueto, David Randell, and Oliver Kosut. Locating and quantifying gas emission sources using remotely obtained concentration data. *Atmospheric Environment*, 74:141–158, August 2013.

[81] Ekkehard Holzbecher. *Environmental Modeling*. Springer, 2nd edition, 2012.

[82] Frédéric Hourdin and Olivier Talagrand. Eulerian backtracking of atmospheric tracers. I: Adjoint derivation and parametrization of subgrid-scale transport. *Quarterly Journal of the Royal Meteorological Society*, 132(615):567–583, January 2006.

[83] Marco Huber. *Probabilistic Framework for Sensor Management*. PhD thesis, Universität Karlsruhe (TH), April 2009.

[84] Marco Huber, Dietrich Brunn, and Uwe D. Hanebeck. Closed-Form Prediction of Nonlinear Dynamic Systems by Means of Gaussian Mixture Approximation of the Transition Density. In *Proceedings of the 2006 IEEE International Conference on Multisensor Fusion and Integration for Intelligent Systems (MFI)*, pages 98–103, Heidelberg, Germany, September 2006.

[85] Marco F. Huber. Recursive Gaussian Process Regression. In *Proceedings of the 38th International Conference on Acoustics, Sound, and Signal Processing (ICASSP)*, pages 3362–3366, Vancouver, BC, Canada, May 2013.

[86] Marco F. Huber, Tim Bailey, Hugh Durrant-Whyte, and Uwe D. Hanebeck. On Entropy Approximation for Gaussian Mixture Random Vectors. In *Proceedings of the 2008 IEEE International Confer-*

ence on Multisensor Fusion and Integration for Intelligent Systems (MFI), pages 181–188, Seoul, Republic of Korea, August 2008.

[87] Marco F. Huber, Dietrich Brunn, and Uwe D. Hanebeck. Efficient Nonlinear Measurement Updating based on Gaussian Mixture Approximation of Conditional Densities. In *Proceedings of the 2007 American Control Conference (ACC)*, pages 4425–4430, New York, New York, July 2007.

[88] Marco F. Huber, Tobias Dencker, Masoud Roschani, and Jürgen Beyerer. Bayesian Active Object Recognition via Gaussian Process Regression. In *Proceedings of the 15th International Conference on Information Fusion (Fusion)*, July 2012.

[89] Marco F. Huber and Uwe D. Hanebeck. Gaussian Filter based on Deterministic Sampling for High Quality Nonlinear Estimation. In *Proceedings of the 17th IFAC World Congress*, pages 13527–13532, Seoul, Republic of Korea, July 2008.

[90] Marco F. Huber and Uwe D. Hanebeck. Progressive Gaussian Mixture Reduction. In *Proceedings of the 11th International Conference on Information Fusion (Fusion)*, Cologne, Germany, July 2008.

[91] Kazufumi Ito and Kaiqi Xiong. Gaussian Filters for Nonlinear Filtering Problems. *IEEE Transactions on Automatic Control*, 45(5):910–927, May 2000.

[92] Edwin T. Jaynes. *Probability Theory*. Cambridge University Press, 2003.

[93] Andrew H. Jazwinski. *Stochastic Processes and Filtering Theory*. Dover Publications, Inc., 2007.

[94] Simon Julier, Jeffrey Uhlmann, and Hugh F. Durrant-Whyte. A New Method for the Nonlinear Transformation of Means and Covariances in Filters and Estimators. *IEEE Transactions on Automatic Control*, 45(3):477–482, 2000.

[95] Simon J. Julier and Jeffrey K. Uhlmann. A New Extension of the Kalman Filter to Nonlinear Systems. In *International Symposium on Aerospace/Defence Sensing, Simulation and Control*, 1997.

[96] Simon J. Julier and Jeffrey K. Uhlmann. Unscented Filtering and Nonlinear Estimation. *Proceedings of the IEEE*, 92(3):401–422, 2004.

[97] Rudolf E. Kalman. A new Approach to Linear Filtering and Prediction Problems. *Transactions of the ASME, Journal of Basic Engineering*, 82 (Series D)(1):35–45, 1960.

[98] Raymond Kan. From Moments of Sum to Moments of Product. *Journal of Multivariate Analysis*, 99(3):542–554, March 2008.

[99] Elliott D. Kaplan and Christopher Hegarty. *Understanding GPS: Principles and Applications*. Artech House, 2nd edition, 2005.

[100] Uwe Kiencke and Ralf Eger. *Meßtechnik*. Springer, 6th edition, 2001.

[101] Kiseon Kim and Georgy Shevlyakov. Why Gaussianity? *IEEE Signal Processing Magazine*, pages 102–113, March 2008.

[102] Genshiro Kitagawa. Monte Carlo Filter and Smoother for Non-Gaussian Nonlinear State Space Models. *Journal of Computational and Graphical Statistics*, 5(1):1–25, 1996.

[103] Vesa Klumpp, Frederik Beutler, Uwe D. Hanebeck, and Dietrich Fränken. The Sliced Gaussian Mixture Filter with Adaptive State Decomposition Depending on Linearization Error. In *Proceedings of the 13th International Conference on Information Fusion*, Edinburgh, United Kingdom, July 2010.

[104] Jonathan Ko and Dieter Fox. GP-BayesFilters: Bayesian Filtering using Gaussian Process Prediction and Observation Models. *Autonomous Robots*, 27(1):75–90, 2009.

[105] Jonathan Ko, Daniel J. Klein, Dieter Fox, and Dirk Haehnel. GP-UKF: Unscented Kalman Filters with Gaussian Process Prediction and Observation Models. In *Proceedings of the 2007 IEEE/RSJ International Conference on Intelligent Robots and Systems*, pages 1901–1907, San Diego, CA, October-November 2007.

[106] Jayesh H. Kotecha and Petar M. Djurić. Gaussian Particle Filtering. *IEEE Transactions on Signal Processing*, 51(10):2592–2601, 2003.

[107] Jayesh H. Kotecha and Petar M. Djurić. Gaussian Sum Particle Filtering. *IEEE Transactions on Signal Processing*, 51(10):2602–2612, 2003.

[108] Edgar Kraft. A Quaternion-Based Unscented Kalman Filter for Orientation Tracking. In *Proceedings of the Sixth International Conference of Information Fusion*, volume 1, pages 47–54, 2003.

[109] Stuart C. Kramer and Harold W. Sorenson. Recursive Bayesian estimation using piece-wise constant approximations. *Automatica*, 24(6):789–801, November 1988.

[110] Andreas Krause and Carlos Guestrin. Nonmyopic Active Learning of Gaussian Processes: An Exploration-Exploitation Approach. In *Proceedings of the 24 th International Conference on Machine Learning*, pages 449–456, Corvallis, OR, 2007.

[111] Peter Krauthausen. *Learning Dynamic Systems for Intention Recognition in Human-Robot-Cooperation*. PhD thesis, Karlsruhe Institute of Technology (KIT), 2012.

[112] Catherine Laporte and Tal Arbel. Efficient Discriminant Viewpoint Selection for Active Bayesian Recognition. *International Journal of Computer Vision*, 68:267–287, July 2006.

[113] Tine Lefebvre, Herman Bruyninckx, and Joris De Schutter. Comments on "A New Method for the Nonlinear Transformation of Means and Covariances in Filters and Estimators". *IEEE Transactions on Automatic Control*, 45(8):1406–1408, 2002.

[114] Tine Lefebvre, Herman Bruyninckx, and Joris De Schutter. *Nonlinear Kalman Filtering for Force-Controlled Robot Tasks.* Springer Berlin, 2005.

[115] Jun S. Liu and Rong Chen. Blind Deconvolution via Sequential Imputations. *Journal of the American Statistical Association,* 90(430):567–576, June 1995.

[116] David G. Lowe. Object recognition from local scale-invariant features. In *Proceedings of the 7th International Conference on Computer Vision,* volume 2, pages 1150–1157, Kerkyra, Greece, September 1999.

[117] Mihai Bogdan Luca, Stéphane Azou, Gilles Burel, and Alexandru Serbanescu. On Exact Kalman Filtering of Polynomial Systems. *IEEE Transactions on Circuits and Systems—I: Regular Papers,* 53(6):1329–1340, June 2006.

[118] David J. C. MacKay. Comparison of Approximate Methods for Handling Hyperparameters. *Neural Computation,* 11(5):1035–1068, 1999.

[119] Dimitris E. Manolakis. Efficient Solution and Performance Analysis of 3-D Position Estimation by Trilateration. *IEEE Transactions on Aerospace and Electronic Systems,* 32(4):1239–1248, October 1996.

[120] John C. Mason and David C. Handscomb. *Chebyshev Polynomials.* Chapman & Hall/CRC, 2003.

[121] Peter Maybeck. *Stochastic Models, Estimation, and Control, Volume 2.* Academic Press, 1982.

[122] Vladimir Maz'ya and Gunther Schmidt. On approximate approximations using gaussian kernels. *IMA Journal of Numerical Analysis,* 16:13–29, 1996.

[123] Leonard A. McGee and Stanley F. Schmidt. Discovery of the Kalman Filter as a Practical Tool for Aerospace and Industry. Technical memorandum 86847, NASA, 1985.

[124] H. S. Migon and P. J. Harrison. An application of non-linear Bayesian forecasting to television advertising. In J. M. Bernardo, M. H DeGroot, D. V. Lindley, and A. F. M. Smith, editors, *Bayesian Statistics 2.* Valencia University Press, 1985.

[125] Javier G. Monroya, Achim J. Lilienthalc, Jose-Luis Blancob, Javier Gonzalez-Jimeneza, and Marco Trincavelli. Probabilistic Gas Quantification with MOX Sensors in Open Sampling Systems - A Gaussian Process Approach. *Sensors and Actuators B: Chemical,* 188:298–312, November 2013.

[126] Mark R. Morelande and Bill Moran. An Unscented Transformation for Conditionally Linear Models. In *IEEE International Conference on Acoustics, Speech and Signal Processing (ICASSP),* pages III–1417–III–1420, April 2007.

[127] Hiroshi Murase and Shree K. Nayar. Visual learning and recognition of 3-D objects from appearance. *International Journal Computer Vision,* 14:5–24, January 1995.

[128] Kevin P. Murphy. *Machine Learning: A Probabilistic Perspective.* MIT Press, 2012.

[129] Christian Musso, Nadia Oudjane, and Francois Le Gland. Improving Regularised Particle Filters. In A. Doucet, N. de Freitas, and N. Gordon, editors, *Sequential Monte Carlo Methods in Practice,* chapter 12. New York: Springer Verlag, 2001.

[130] A. Nemra and N. Aouf. Robust INS/GPS Sensor Fusion for UAV Localization Using SDRE Nonlinear Filtering. *IEEE Sensors Journal,* 10(4):789–798, April 2010.

[131] Duy Nguyen-Tuong and Jan Peters. Local Gaussian Processes Regression for Real-time Model-based Robot Control. In *Proceeding of the IEEE/RSJ International Conference on Intelligent Robots and Systems (IROS)*, pages 380–385, 2008.

[132] Duy Nguyen-Tuong, Matthias Seeger, and Jan Peters. Real-Time Local GP Model Learning. In Olivier Sigaud and Jan Peters, editors, *From Motor Learning to Interaction Learning in Robots*, volume 264 of *Studies in Computational Intelligence*, pages 193–207. Springer-Verlag, 2010.

[133] Daniel Nikovski and Matthew Brand. Non-Linear Stochastic Control in Continuous State Spaces by Exact Integration in Bellman's Equations. In *Proceedings of the 2003 International Conference on Automated Planning and Scheduling*, pages 91–95, 2003.

[134] Magnus Nørgaard, Niels K. Poulsen, and Ole Ravn. New Developments in State Estimation for Nonlinear Systems. *Automatica*, 36(11):1627–1638, November 2000.

[135] Lucas Paletta and Axel Pinz. Active Object Recognition By View Integration and Reinforcement Learning. *Robotics and Autonomous Systems*, 31:71–86, 2000.

[136] Lucy Y. Pao. Multisensor Multitarget Mixture Reduction Algorithms for Tracking. *AIAA Journal of Guidance, Control and Dynamics*, 17:1205–1211, 1994.

[137] Kalyanapuram R. Parthasarathy. *Probability Measures on Metric Spaces*. American Mathematical Society, new edition, 2005.

[138] Frank Pasquill. The estimation of the dispersion of windborne material. *The Meteorological Magazine*, 90(1063):33–49, 1961.

[139] Fernando Pérez-Cruz, Steven Van Vaerenbergh, Juan José Murillo-Fuentes, Miguel Lázao-Gredilla, and Ignacio Santamaría. Gaussian Processes for Nonlinear Signal Processing. *IEEE Signal Processing Magazine*, 30(4):40–50, July 2013.

[140] William H. Press, Saul A. Teukolsky, William T. Vetterling, and Brian P. Flannery. *Numerical Recipes: The Art of Scientific Computing*, chapter 17: Integration of Ordinary Differential Equations. Cambridge University Press, 3rd edition, 2007.

[141] Joaquin Quiñonero-Candela and Carl E. Rasmussen. A Unifying View of Sparse Approximate Gaussian Process Regression. *Journal of Machine Learning Research*, 6:1939–1959, 2005.

[142] Ananth Ranganathan, Ming-Hsuan Yang, and Jeffrey Ho. Online Sparse Gaussian Process Regression and Its Applications. *IEEE Transactions on Image Processing*, 20(2):391–404, February 2011.

[143] K. Shankar Rao. Source estimation methods for atmospheric dispersion. *Atmospheric Environment*, 41:6964–6973, 2007.

[144] Carl E. Rasmussen and Christopher K. I. Williams. *Gaussian Processes for Machine Learning*. The MIT Press, 2006.

[145] H. E. Rauch, F. Tung, and C. T. Striebel. Maximum Likelihood Estimates of Linear Dynamic Systems. *AIAA Journal*, 3(8):1445–1450, August 1965.

[146] Andreas Rauh, Kai Briechle, and Uwe D. Hanebeck. Nonlinear Measurement Update and Prediction: Prior Density Splitting Mixture Estimator. In *Proceedings of the 2009 IEEE International Conference on Control Applications (CCA)*, July 2009.

[147] Andreas Rauh and Uwe D. Hanebeck. Calculating Moments of Exponential Densities Using Differential Algebraic Equations. *IEEE Signal Processing Letters*, 10(5):144–147, May 2003.

[148] Andreas Rauh and Uwe D. Hanebeck. Moment-Based Prediction Step for Nonlinear Discrete-Time Dynamic Systems Using Exponential Densities. In *Proceedings of the 44th IEEE Conference on Decision and Control and European Control Conference*, pages 1923–1928, Sevilla, Spain, December 2005.

[149] Steven Reece and Stephen Roberts. An Introduction to Gaussian Processes for the Kalman Filter Expert. In *Proceedings of the 13th International Conference on Information Fusion*, Edinburgh, UK, 2010.

[150] Robert H. Risch. The problem of integration in finite terms. *Transactions of the American Mathematical Society*, 139:167–189, May 1969.

[151] Robert H. Risch. The solution of the problem of integration in finite terms. *Bulletin of the American Mathematical Society*, 76:605–608, 1970.

[152] Theodore Rivlin. *Chebyshev Polynomials.* New York: Wiley, 1990.

[153] Patrick Rößler, Frederik Beutler, Uwe D. Hanebeck, and Norbert Nitzsche. Motion Compression Applied to Guidance of a Mobile Teleoperator. In *Proceedings of the 2005 IEEE International Conference on Intelligent Robots and Systems (IROS)*, pages 2495–2500, 2005.

[154] Sumantra Dutta Roy, Santanu Chaudhury, and Subhashis Banerjee. Active recognition through next view planning: a survey. *Pattern Recognition*, 37(3):429–446, March 2004.

[155] Alison Rudd, Alan G. Robins, Jason J. Lepley, and Stephen E. Belcher. An Inverse Method for Determining Source Characteristics for Emergency Response Applications. *Boundary-Layer Meteorology*, 144(1):1–20, July 2012.

[156] Andrew R. Runnalls. Kullback-Leibler Approach to Gaussian Mixture Reduction. *IEEE Transactions on Aerospace and Electronic Systems*, 43(3):989–999, July 2007.

[157] Sartaj Sahni. Computationally Related Problems. *SIAM Journal on Computing*, 3(4):262–279, 1974.

[158] David J. Salmond. Mixture reduction algorithms for target tracking in clutter. In *Proceedings of SPIE Signal and Data Processing of Small Targets*, volume 1305, pages 434–445, October 1990.

[159] Simo Särkkä. Unscented rauch-tung-striebel smoother. *IEEE Transactions on Automatic Control*, 53(3):845–849, 2008.

[160] Simo Särkkä. *Bayesian Filtering and Smoothing*. Cambridge University Press, 2013.

[161] Juha Sarmavuori and Simo Särkkä. Fourier-Hermite Kalman Filter. *IEEE Transactions on Automatic Control*, 57(6):1511–1515, June 2012.

[162] Tor S. Schei. A finite difference method for linearizing in nonlinear estimation algorithms. *Automatica*, 33(11):2051–2058, November 1997.

[163] Dennis Schieferdecker and Marco F. Huber. Gaussian Mixture Reduction via Clustering. In *Proceedings of the 12th International Conference on Information Fusion (Fusion)*, pages 1536–1543, Seattle, Washington, July 2009.

[164] Bernhard Schölkopf and Alexander Smola. *Learning with Kernels: Support Vector Machines, Regularization, Optimization, and Beyond*. Adaptive computation and machine learning series. MIT Press, Cambridge, Massachusetts, 2002.

[165] Bernhard Schölkopf and Alexander J. Smola. *Learning with Kernels: Support Vector Machines, Regularization, Optimization, and Beyond*. MIT Press, 2002.

[166] Thomas Schön, Fredrik Gustafsson, and Per-Johan Nordlund. Marginalized Particle Filters for Mixed Linear/Nonlinear State-Space Models. *IEEE Transactions on Signal Processing*, 53(7):2279–2287, July 2005.

[167] Fred C. Schweppe. *Uncertain Dynamic Systems.* Prentice–Hall, 1973.

[168] David W. Scott and William F. Szewczyk. From Kernels to Mixtures. *Technometrics,* 43(3):323–335, 2001.

[169] Fernando Seco, Antonio R. Jiménez, Carlos Prieto, Javier Roa, and Katerina Koutsou. A Survey of Mathematical Methods for Indoor Localization. In *IEEE International Symposium on Intelligent Signal Processing,* pages 9–14, August 2009.

[170] Inanc Senocak, Nicolas W. Hengartner, Margaret B. Short, and W. Brent Daniel. Stochastic Event Reconstruction of Atmospheric Contaminant Dispersion Using Bayesian Inference. *Atmospheric Environment,* 42(33):7718–7727, October 2008.

[171] Miroslav Simandl and Jindrich Duník. Sigma point gaussian sum filter design using square root unscented filters. In *Proceedings of the 16th IFAC World Congress,* Prague, Czech Republic, July 2005.

[172] Miroslav Simandl and Jindrich Duník. Design of derivative-free smoothers and predictors. In *Proceedings of the 14th IFAC Symposium on System Identification,* pages 1240–1245, Australia, 2006.

[173] Miroslav Simandl, Jakub Královeca, and Torsten Söderström. Advanced point mass method for nonlinear state estimation. *Automatica,* 42(7):1133–1145, July 2006.

[174] Dan Simon. *Optimal State Estimation: Kalman, H-Infinity, and Nonlinear Approaches.* Wiley & Sons, 1st edition, 2006.

[175] John Skilling. Bayesian Numerical Analysis. In W. T. Grandy, Jr. and P. W. Milonni, editors, *Physics and Probability,* pages 207–222. Cambridge University Press, 1993.

[176] Alex J. Smola and Peter Bartlett. Sparse Greedy Gaussian Process Regression. In *Advances in Neural Information Processing Systems 13,* pages 619–625. MIT Press, 2001.

[177] Ed Snelson and Zoubin Ghahramani. Sparse Gaussian processes using pseudo-inputs. In Y. Weiss, B. Schölkopf, and J. Platt, editors, *Advances in Neural Information Processing Systems 18*, pages 1259–1266. The MIT Press, 2006.

[178] Edward Snelson, Carl Edward Rasmussen, and Zoubin Ghahramani. Warped Gaussian Processes. In *Advances in Neural Information Processing Systems 16*. The MIT Press, 2004.

[179] Michael D. Sohn, Pamela Reynolds, Navtej Singh, and Ashok J. Gadgil. Rapidly Locating and Characterizing Pollutant Releases in Buildings. *Journal of the Air & Waste Management Association*, 52(12):1422–1432, 2002.

[180] Francisco J. Solis and Roger J-B. Wets. Minimization by Random Search Techniques. *Mathematics of Operations Research*, 6(1):19–30, February 1981.

[181] Harold W. Sorenson and Daniel L. Alspach. Recursive bayesian estimation using gaussian sums. *Automatic*, 7(4):465–479, July 1971.

[182] Harold W. Sorenson and Allen R. Stubberud. Non-linear filtering by approximation of the a posteriori density. *International Journal of Control*, 8(1):33–51, 1968.

[183] Krishnaswamy Srinivasan. State Estimation by Orthogonal Expansion of Probability Distributions. *IEEE Transactions on Automatic Control*, 15(1):3–10, February 1970.

[184] Rajan Srinivasan. *Importance Sampling: Applications in Communications and Detection*. Springer, 2002.

[185] John M. Stockie. The Mathematics of Atmospheric Dispersion Modelling. *SIAM Review*, 53(2):349–372, 2011.

[186] Lawrence D. Stone, Roy L. Streit, Thomas L. Corwin, and Kristine L. Bell. *Bayesian Multiple Target Tracking.* Artech House, 2nd edition, 2014.

[187] Volker Strassen. Gaussian Elimination is not Optimal. *Numerische Mathematik,* 13(4):354–356, August 1969.

[188] Jürgen Sturm. *Approaches to Probabilistic Model Learning for Mobile Manipulation Robots.* PhD thesis, Albert-Ludwigs-Universität Freiburg im Breisgau, 2011.

[189] Hsi Guang Sung. *Gaussian Mixture Regression and Classification.* PhD thesis, Rice University, May 2004.

[190] Richard Szeliski. *Computer Vision: Algorithms and Applications,* chapter 14 – Recognition. Springer London, 2010.

[191] Nassim N. Taleb. *The Black Swan: The Impact of the Highly Improbable.* Random House Trade, 2nd edition, 2010.

[192] Wing Ip Tam, K.N. Plataniotis, and D. Hatzinakos. An adaptive Gaussian sum algorithm for radar tracking. *Signal Processing,* 77:85–104, 1999.

[193] D. Tenne and T. Singh. The Higher Order Unscented Filter. In *Proceedings of the American Control Conference,* pages 2441–2446, June 2003.

[194] Gabriel Terejanu, Puneet Singla, Tarunraj Singh, and Peter D. Scott. A Novel Gaussian Sum Filter Method for Accurate Solution to Nonlinear Filtering Problem. In *Proceedings of the 11th International Conference on Information Fusion,* 2008.

[195] Gabriel Terejanu, Puneet Singla, Tarunraj Singh, and Peter D. Scott. Uncertainty Propagation for Nonlinear Dynamic Systems Using Gaussian Mixture Models. *Journal of Guidance, Control, and Dynamics,* 31(6):1623–1633, November–December 2008.

[196] Federico Thomas and Lluís Ros. Revisiting trilateration for robot localization. *IEEE Transactions on Robotics*, 21:93–101, 2005.

[197] Sebastian Thrun, Wolfram Burgard, and Dieter Fox. *Probabilistic Robotics.* The MIT Press, 2005.

[198] Petr Tichavský, Carlos H. Muravchik, and Arye Nehorai. Posterior Cramér-Rao Bounds for Discrete-Time Nonlinear Filtering. *IEEE Transactions on Signal Processing*, 46(5):1386–1396, May 1998.

[199] Volker Tresp. A Bayesian Committee Machine. *Neural Computation*, 12(11):2719–2741, 2000.

[200] Raquel Urtasun and Trevor Darrell. Discriminative Gaussian Process Latent Variable Model for Classification. In *Proceedings ot the 24th International Conference on Machine Learning*, Corvallis, OR, 2007.

[201] Rudolph van der Merwe, Arnaud Doucet, Nando de Freitas, and Eric Wan. The Unscented Particle Filter. In *Advances in Neural Information Processing Systems*, volume 12, pages 666–672. MIT Press, 2000.

[202] Nuno Vasconcelos. Image Indexing with Mixture Hierarchies. In *Proceedings of the IEEE Conference in Computer Vision and Pattern Recognition*, 2001.

[203] Shrihari Vasudevan, Fabio Ramos, Eric Nettleton, and Hugh Durrant-Whyte. Gaussian Process Modeling of Large-Scale Terrain. *Journal of Field Robotics*, 26(10):812–840, 2009.

[204] Tom Vercauteren and Xiaodong Wang. Decentralized Sigma-Point Information Filters for Target Tracking in Collaborative Sensor Networks. *IEEE Transactions on Signal Processing*, 53(8):2997–3009, August 2005.

[205] Grace Wahba, Xiwu Lin, Fangyu Gao, Dong Xiang, Ronald Klein, and Barbara Klein. The Bias-Variance Tradeoff and the Randomized GACV. In *Advances in Neural Information Processing Systems 11*, pages 620–626. MIT Press, 1999.

[206] Eric A. Wan and Rudolph van der Merwe. The Unscented Kalman Filter. In Simon Haykin, editor, *Kalman Filtering and Neural Networks*, chapter The Unscented Kalman Filter, pages 221–280. John Wiley & Sons, Inc., 2001.

[207] Greg Welch, B. Danette Allen, Adrian Ilie, and Gary Bishop. Measurement Sample Time Optimization for Human Motion Tracking/-Capture Systems. In *Proceedings of Trends and Issues in Tracking for Virtual Environments, Workshop at the IEEE Virtual Reality 2007 Conference*, 2007.

[208] Mike West. Approximating Posterior Distributions by Mixtures. *Journal of the Royal Statistical Society: Series B*, 55(2):409–422, 1993.

[209] Mike West and Jeff Harrison. *Bayesian Forecasting and Dynamic Models*, chapter 14: Exponential Family Dynamic Models, pages 534–555. Springer, 1997.

[210] Jason L. Williams. *Information Theoretic Sensor Management*. PhD thesis, Massachusetts Institute of Technology, February 2007.

[211] Jason L. Williams and Peter S. Maybeck. Cost-Function-Based Hypothesis Control Techniques for Multiple Hypothesis Tracking. In *Proceedings of SPIE Signal and Data Processing of Small Targets*, volume 5428, April, 2004.

[212] W. Murray Wonham. Some Applications of Stochastic Differential Equations to Optimal Nonlinear Filtering. *Journal of the Society for Industrial and Applied Mathematics, Series A: Control*, 2(3):347–369, 1964.

[213] Sally A. Wood, Wenxin Jian, and Martin Tanner. Bayesian mixture of splines for spatially adaptive nonparameteric regression. *Biometrika*, 89(3):513–528, 2002.

[214] Yuanxin Wu, Dewen Hu, Meiping Wu, and Xiaoping Hu. A Numerical-Integration Perspective on Gaussian Filters. *IEEE Transactions on Signal Processing*, 54(8):2910–2921, August 2006.

[215] Renato Zanetti. Recursive Update Filtering for Nonlinear Estimation. *IEEE Transactions on Automatic Control*, 57(6):1481–1490, June 2012.

[216] Arnold Zellner. Optimal Information Processing and Bayes's Theorem. *The American Statistician*, 42(4):278–280, 1988.

[217] Yong Zhang and Li Wang. Particle Filtering Method for Source Localization in Wireless Sensor Network. In *Advanced Technology in Teaching: Selected papers from the 2012 International Conference on Teaching and Computational Science (ICTCS 2012)*, volume 163, pages 517–523. Springer, 2013.

Part II

Publications

Paper A

Gaussian Filtering using State Decomposition Methods

Authors:

Frederik Beutler, Marco F. Huber, and Uwe D. Hanebeck

Published in:

Proceedings of the 12th International Conference on Information Fusion (Fusion), pages 579–586, Seattle, WA, USA, July 2009.

Link to:

`http://isif.org/fusion/proceedings/fusion09CD/data/papers/0371.pdf`

Abstract:

State estimation for nonlinear systems generally requires approximations of the system or the probability densities, as the occurring prediction and filtering equations cannot be solved in closed form.

For instance, Linear Regression Kalman Filters like the Unscented Kalman Filter or the considered *Gaussian Filter* propagate a small set of sample points through the system to approximate the posterior mean and covariance matrix. To reduce the number of sample points, special structures of the system and measurement equation can be taken into account. In this paper, two principles of system decomposition are considered and applied to the Gaussian Filter. One principle exploits that only a part of the state vector is directly observed by the measurement. The second principle separates the system equations into linear and nonlinear parts in order to merely approximate the nonlinear part of the state. The benefits of both decompositions are demonstrated on a real-world example.

Paper B

Semi-Analytic Gaussian Assumed Density Filter

Authors:

Marco F. Huber, Frederik Beutler, and Uwe D. Hanebeck

Published in:

Proceedings of the 2011 American Control Conference (ACC), pages 3006–3011, San Francisco, CA, USA, June 2011.

Link to:

http://dx.doi.org/10.1109/ACC.2011.5991332

Abstract:

For Gaussian Assumed Density Filtering based on moment matching, a framework for the efficient calculation of posterior moments is proposed that exploits the structure of the given nonlinear system. The key idea is a careful discretization of some dimensions of the state space only in order to decompose the system into a set of nonlinear subsystems that are conditionally integrable in closed form. This approach is more efficient than full discretization approaches. In addition, the new decomposition is far more general than known Rao-Blackwellization approaches relying on conditionally linear subsystems. As a result, the new framework is applicable to a much larger class of nonlinear systems.

Paper C

Chebyshev Polynomial Kalman Filter

Authors:

Marco F. Huber

Published in:

Digital Signal Processing, vol. 23, no. 5, pages 1620–1629, September 2013.

Link to:

`http://dx.doi.org/10.1016/j.dsp.2013.06.005`

Abstract:

A novel Gaussian state estimator named Chebyshev Polynomial Kalman Filter is proposed that exploits the exact and closed-form calculation of posterior moments for polynomial nonlinearities. An arbitrary nonlinear system is at first approximated via a Chebyshev polynomial series. By exploiting special properties of the Chebyshev polynomials, exact expressions for mean and variance are then provided in computationally efficient vector-matrix notation for prediction and measurement update. Approximation and state estimation are performed in a black-box fashion without the need of manual operation or manual inspection. The superior performance of the Chebyshev Polynomial Kalman Filter compared to state-of-the-art Gaussian estimators is demonstrated by means of numerical simulations and a real-world application.

Paper D

Gaussian Filtering for Polynomial Systems Based on Moment Homotopy

Authors:

Marco F. Huber and Uwe D. Hanebeck

Published in:

Proceedings to the 16th International Conference on Information Fusion (Fusion), pages 1080–1087, Istanbul, Turkey, July 2013.

Link to:

`http://ieeexplore.ieee.org/xpl/articleDetails.jsp?tp=&arnumber=6641116`

Abstract:

This paper proposes Gaussian filters for polynomial systems with efficient solutions for both the prediction and the filter step. For the prediction step, computationally efficient closed-form solutions are derived for calculating the exact moments. In order to achieve a higher estimation quality, the filter step is solved without the usual additional assumption that state and measurement are jointly Gaussian distributed. As this significantly complicates the required moment calculation, a homotopy continuation method is employed that yields almost optimal results.

Paper E

(Semi-)Analytic Gaussian Mixture Filter

Authors:

Marco F. Huber, Frederik Beutler, and Uwe D. Hanebeck

Published in:

Proceedings of the 18th IFAC World Congress, pages 10014–10020, Milano, Italy, August 2011.

Link to:

`http://dx.doi.org/10.3182/20110828-6-IT-1002.03359`

Abstract:

In nonlinear filtering, special types of Gaussian mixture filters are a straightforward extension of Gaussian filters, where linearizing the system model is performed individually for each Gaussian component. In this paper, two novel types of linearization are combined with Gaussian mixture filters. The first linearization is called analytic stochastic linearization, where the linearization is performed analytically and exactly, i.e., without Taylor-series expansion or approximate sample-based density representation. In cases where a full analytical linearization is not possible, the second approach decomposes the nonlinear system into a set of nonlinear subsystems that are conditionally integrable in closed form. These approaches are more accurate than fully applying classical linearization.

Paper F

Adaptive Gaussian Mixture Filter

Authors:

Marco F. Huber

Published in:

Proceedings of the 14th International Conference on Information Fusion (Fusion), Chicago, Illinois, July 2011.

Link to:

`http://isif.org/fusion/proceedings/Fusion_2011/data/papers/075.pdf`

Abstract:

Gaussian mixtures are a common density representation in nonlinear, non-Gaussian Bayesian state estimation. Selecting an appropriate number of Gaussian components, however, is difficult as one has to trade of computational complexity against estimation accuracy. In this paper, an adaptive Gaussian mixture filter based on statistical linearization is proposed. Depending on the nonlinearity of the considered estimation problem, this filter dynamically increases the number of components via splitting. For this purpose, a measure is introduced that allows for quantifying the locally induced linearization error at each Gaussian mixture component. The deviation between the nonlinear and the linearized state space model is evaluated for determining the splitting direction. The proposed approach is not restricted to a specific statistical linearization method. Simulations show the superior estimation performance compared to related approaches and common filtering algorithms.

Paper G

Superficial Gaussian Mixture Reduction

Authors:

Marco F. Huber, Peter Krauthausen, and Uwe D. Hanebeck

Published in:

INFORMATIK 2011 - the 41th Annual Conference of the Gesellschaft für Informatik e.V. (GI), 6th Workshop Sensor Data Fusion: Trends, Solutions, Applications (SDF), Berlin, Germany, October 2011.

Link to:

`https://www.user.tu-berlin.de/komm/CD/paper/100147.pdf`

Abstract:

Many information fusion tasks involve the processing of Gaussian mixtures with simple underlying shape, but many components.

This paper addresses the problem of reducing the number of components, allowing for faster density processing. The proposed approach is based on identifying components irrelevant for the overall density's shape by means of the curvature of the density's surface. The key idea is to minimize an upper bound of the curvature while maintaining a low global reduction error by optimizing the weights of the original Gaussian mixture only. The mixture is reduced by assigning zero weights to reducible components. The main advantages are an alleviation of the model selection problem, as the number of components is chosen by the algorithm automatically, the derivation of simple curvature-based penalty terms, and an easy, efficient implementation. A series of experiments shows the approach to provide a good trade-off between quality and sparsity.

Paper H

Analytic Moment-based Gaussian Process Filtering

Authors:

Marc P. Deisenroth, Marco F. Huber, and Uwe D. Hanebeck

Published in:

Proceedings of the 26th International Conference on Machine Learning (ICML), Montreal, Canada, June 2009.

Link to:

`http://dx.doi.org/10.1145/1553374.1553403`

Abstract:

We propose an analytic moment-based filter for nonlinear stochastic dynamic systems modeled by Gaussian processes. Exact expressions for the expected value and the covariance matrix are provided for both the prediction step and the filter step, where an additional Gaussian assumption is exploited in the latter case. Our filter does not require further approximations. In particular, it avoids finite-sample approximations. We compare the filter to a variety of Gaussian filters, that is, the EKF, the UKF, and the recent GP-UKF proposed by [1].

References

[1] Jonathan Ko, Daniel J. Klein, Dieter Fox, and Dirk Haehnel. Gaussian Processes and Reinforcement Learning for Identification and Control of an Autonomous Blimp. In *Proceedings of the International Conference on Robotics and Automation (ICRA)*, pages 742–747, Rome, Italy, April 2007.

Paper I

Robust Filtering and Smoothing with Gaussian Processes

Authors:

Marc P. Deisenroth, Ryan D. Turner, Marco F. Huber, Uwe D. Hanebeck, and Carl E. Rasmussen

Published in:

IEEE Transactions on Automatic Control, vol. 57, no. 7, pages 1865–1871, July 2012.

Link to:

`http://dx.doi.org/10.1109/TAC.2011.2179426`

Abstract:

We propose a principled algorithm for robust Bayesian filtering and smoothing in nonlinear stochastic dynamic systems when both the transition function and the measurement function are described by non-parametric Gaussian process (GP) models. GPs are gaining increasing importance in signal processing, machine learning, robotics, and control for representing unknown system functions by posterior probability distributions. This modern way of "system identification" is more robust than finding point estimates of a parametric function representation. In this article, we present a principled algorithm for robust analytic smoothing in GP dynamic systems, which are increasingly used in robotics and control. Our numerical evaluations demonstrate the robustness of the proposed approach in situations where other state-of-the-art Gaussian filters and smoothers can fail.

Paper J

Recursive Gaussian Process Regression

Authors:

Marco F. Huber

Published in:

Proceedings of the 38th International Conference on Acoustics, Sound, and Signal Processing (ICASSP), pages 3362–3366, Vancouver, BC, Canada, May 2013.

Link to:

`http://dx.doi.org/10.1109/ICASSP.2013.6638281`

Abstract:

For large data sets, performing Gaussian process regression is computationally demanding or even intractable. If data can be processed sequentially, the recursive regression method proposed in this paper allows incorporating new data with constant computation time. For this purpose two operations are performed alternating on a fixed set of so-called basis vectors used for estimating the latent function: First, inference of the latent function at the new inputs. Second, utilization of the new data for updating the estimate. Numerical simulations show that the proposed approach significantly reduces the computation time and at the same time provides more accurate estimates compared to existing on-line and/or sparse Gaussian process regression approaches.

Paper K

Recursive Gaussian Process: On-line Regression and Learning

Authors:

Marco F. Huber

Published in:

Pattern Recognition Letters, vol. 45, pages 85–91, August 2014.

Link to:

`http://dx.doi.org/10.1016/j.patrec.2014.03.004`

Abstract:

Two approaches for on-line Gaussian process regression with low computational and memory demands are proposed. The first approach assumes known hyperparameters and performs regression on a set of basis vectors that stores mean and covariance estimates of the latent function. The second approach additionally learns the hyperparameters on-line. For this purpose, techniques from nonlinear Gaussian state estimation are exploited. The proposed approaches are compared to state-of-the-art sparse Gaussian process algorithms.

Paper L

Optimal Stochastic Linearization for Range-Based Localization

Authors:

Frederik Beutler, Marco F. Huber, and Uwe D. Hanebeck

Published in:

Proceedings of the 2010 IEEE/RSJ International Conference on Intelligent Robots and Systems (IROS), pages 5731–5736, Taipei, Taiwan, October 2010.

Link to:

`http://dx.doi.org/10.1109/IROS.2010.5649076`

Abstract:

In range-based localization, the trajectory of a mobile object is estimated based on noisy range measurements between the object and known landmarks. In order to deal with this uncertain information, a Bayesian state estimator is presented, which exploits optimal stochastic linearization. Compared to standard state estimators like the Extended or Unscented Kalman Filter, where a point-based Gaussian approximation is used, the proposed approach considers the entire Gaussian density for linearization. By employing the common assumption that the state and measurements are jointly Gaussian, the linearization can be calculated in closed form and thus analytic expressions for the range-based localization problem can be derived.

Paper M

Semi-Analytic Stochastic Linearization for Range-Based Pose Tracking

Authors:

Frederik Beutler, Marco F. Huber, and Uwe D. Hanebeck

Published in:

Proceedings of the 2010 IEEE International Conference on Multisensor Fusion and Integration for Intelligent Systems (MFI), pages 44–49, Salt Lake City, UT, USA, September 2010.

Link to:

`http://dx.doi.org/10.1109/MFI.2010.5604468`

Abstract:

In range-based pose tracking, the translation and rotation of an object with respect to a global coordinate system has to be estimated. The ranges are measured between the target and the global frame. In this paper, an intelligent decomposition is introduced in order to reduce the computational effort for pose tracking. Usually, decomposition procedures only exploit conditionally linear models. In this paper, this principle is generalized to conditionally integrable substructures and applied to pose tracking. Due to a modified measurement equation, parts of the problem can even be solved analytically.

Paper N

On-line Dispersion Source Estimation using Adaptive Gaussian Mixture Filter

Authors:

Marco F. Huber

Published in:

Proceedings of the 19th IFAC World Congress, pages 1059–1066, Cape Town, South Africa, August 2014.

Link to:

http://dx.doi.org/10.3182/20140824-6-ZA-1003.00795

Abstract:

The reconstruction of environmental events has gained increased interest in the recent years. In this paper, the focus is on estimating the location and strength of a gas release from distributed measurements. The estimation is formulated as Bayesian inverse problem, which utilizes a Gaussian plume forward model. A novel recursive estimation algorithm based on statistical linearization and Gaussian mixture densities with adaptive component number selection is used in order to allow at the same time accurate and computationally efficient source estimation. The proposed solution is compared against state-of-the-art methods via simulations and a real-word experiment.

Paper O

Bayesian Active Object Recognition via Gaussian Process Regression

Authors:

Marco F. Huber, Tobias Dencker, Masoud Roschani, and Jürgen Beyerer

Published in:

Proceedings of the 15th International Conference on Information Fusion (Fusion), pages 1718–1725, Singapore, July 2012.

Link to:

http://isif.org/fusion/proceedings/fusion12CD/html/pdf/234_215.pdf

Abstract:

This paper is concerned with a Bayesian approach of actively selecting camera parameters in order to recognize a given object from a finite set of object classes. Gaussian process regression is applied to learn the likelihood of image features given the object classes and camera parameters. In doing so, the object recognition task can be treated as Bayesian state estimation problem. For improving the recognition accuracy and speed, the selection of appropriate camera parameters is formulated as a sequential optimization problem. Mutual information is considered as optimization criterion, which aims at maximizing the information from camera observations or equivalently at minimizing the uncertainty of the state estimate.

Karlsruher Schriftenreihe zur Anthropomatik (ISSN 1863-6489)

Herausgeber: Prof. Dr.-Ing. Jürgen Beyerer

Die Bände sind unter www.ksp.kit.edu als PDF frei verfügbar
oder als Druckausgabe bestellbar.

Band 1 Jürgen Geisler
Leistung des Menschen am Bildschirmarbeitsplatz. 2006
ISBN 3-86644-070-7

Band 2 Elisabeth Peinsipp-Byma
Leistungserhöhung durch Assistenz in interaktiven Systemen zur Szenenanalyse. 2007
ISBN 978-3-86644-149-1

Band 3 Jürgen Geisler, Jürgen Beyerer (Hrsg.)
Mensch-Maschine-Systeme. 2010
ISBN 978-3-86644-457-7

Band 4 Jürgen Beyerer, Marco Huber (Hrsg.)
Proceedings of the 2009 Joint Workshop of Fraunhofer IOSB and Institute for Anthropomatics, Vision and Fusion Laboratory. 2010
ISBN 978-3-86644-469-0

Band 5 Thomas Usländer
Service-oriented design of environmental information systems. 2010
ISBN 978-3-86644-499-7

Band 6 Giulio Milighetti
Multisensorielle diskret-kontinuierliche Überwachung und Regelung humanoider Roboter. 2010
ISBN 978-3-86644-568-0

Band 7 Jürgen Beyerer, Marco Huber (Hrsg.)
Proceedings of the 2010 Joint Workshop of Fraunhofer IOSB and Institute for Anthropomatics, Vision and Fusion Laboratory. 2011
ISBN 978-3-86644-609-0

Band 8 Eduardo Monari
Dynamische Sensorselektion zur auftragsorientierten Objektverfolgung in Kameranetzwerken. 2011
ISBN 978-3-86644-729-5

Band 9 Thomas Bader
Multimodale Interaktion in Multi-Display-Umgebungen. 2011
ISBN 3-86644-760-8

Band 10 Christian Frese
Planung kooperativer Fahrmanöver für kognitive Automobile. 2012
ISBN 978-3-86644-798-1

Band 11 Jürgen Beyerer, Alexey Pak (Hrsg.)
Proceedings of the 2011 Joint Workshop of Fraunhofer IOSB and Institute for Anthropomatics, Vision and Fusion Laboratory. 2012
ISBN 978-3-86644-855-1

Band 12 Miriam Schleipen
Adaptivität und Interoperabilität von Manufacturing Execution Systemen (MES). 2013
ISBN 978-3-86644-955-8

Band 13 Jürgen Beyerer, Alexey Pak (Hrsg.)
Proceedings of the 2012 Joint Workshop of Fraunhofer IOSB and Institute for Anthropomatics, Vision and Fusion Laboratory. 2013
ISBN 978-3-86644-988-6

Band 14 Hauke-Hendrik Vagts
Privatheit und Datenschutz in der intelligenten Überwachung: Ein datenschutzgewährendes System, entworfen nach dem „Privacy by Design" Prinzip. 2013
ISBN 978-3-7315-0041-4

Band 15 Christian Kühnert
Data-driven Methods for Fault Localization in Process Technology. 2013
ISBN 978-3-7315-0098-8

Band 16 Alexander Bauer
Probabilistische Szenenmodelle für die Luftbildauswertung. 2014
ISBN 978-3-7315-0167-1

Band 17 Jürgen Beyerer, Alexey Pak (Hrsg.)
Proceedings of the 2013 Joint Workshop of Fraunhofer IOSB and Institute for Anthropomatics, Vision and Fusion Laboratory. 2014
ISBN 978-3-7315-0212-8

Band 18 Michael Teutsch
Moving Object Detection and Segmentation for Remote Aerial Video Surveillance. 2014
ISBN 978-3-7315-0320-0

Band 19 Marco Huber
Nonlinear Gaussian Filtering: Theory, Algorithms, and Applications. 2015
ISBN 978-3-7315-0338-5